T0176742

Modelling of
Engineering Materials

Modelling of Engineering Materials

C. Lakshmana Rao

Faculty in the Department of Applied Mechanics
Indian Institute of Technology (IIT), Madras

&

Abhijit P. Deshpande

Faculty in the Department of Chemical Engineering
Indian Institute of Technology (IIT), Madras

WILEY

Ane Books Pvt. Ltd.

This edition published in 2014

© (2010) C. Lakshmana Rao and Abhijit P. Deshpande

Published by

Ane Books Pvt. Ltd.

4821 Parwana Bhawan, 1st Floor
24 Ansari Road, Darya Ganj, New Delhi-110 002, India
Tel: +91 (011) 2327 6843-44, 2324 6385
Fax: +91 (011) 2327 6863
e-mail: anebooks@vsnl.net
Website: www.anebooks.com

For

John Wiley & Sons Ltd

The Atrium, Southern Gate
Chichester, West Sussex
PO19 8SQ United Kingdom
Tel : +44 (0)1243 779777
Fax : +44 (0)1243 775878
e-mail : customer@wiley.com
Web : www.wiley.com

For distribution in rest of the world other than the Indian sub-continent

ISBN : 978-1-118-91911-8

Designations used by companies to distinguish their products are often claimed as trademarks. All brand names and product names used in this book are trade names, service marks, trademarks or registered trademarks of their respective owners. The publisher is not associated with any product or vendor mentioned in this book.

Library Congress Cataloging-in-Publication Data

A catalogue record for this book is available from the British Library.

Printed at: Thomson Press, India

DEDICATION

The following close relatives and friends of the authors have left the mortal planes during the course of writing the current book. We dedicate this book to their inspiring memories.

To my grandfather
Krishnarao B. Deshpande
(4.8.1907 – 30.7.2008) – APD.

To my grandmother
Krovi Suryakantam
(15.7.1919 – 8.1.2009) – CLR.

To our friend and colleague
Devanathan Veeraraghavan (Dilip)
(28.9.1958 – 5.2.2009) – CLR & APD.

यस्याऽमतं तस्य मतं, मतं यस्य न वेद सः । (केनोपनिषत्)

Yasyāmatam tasya matam, matam yasya na veda saḥ (Kenopaniṣad)

For those who consider it to be not known, it is known. For those who claim to have known it, it is truly not known.

Preface

Engineers who are designing engineering systems using materials, often need a mathematical model that describes a material response, when a material is subjected to mechanical, thermal, electrical or other fields. Continuum mechanics attempts to provide the necessary mathematical framework that is useful in predicting the material response. *Continuum mechanics* is a classical as well as an emerging field that is exceedingly relevant to many researchers and practicing engineers in the fields of mechanical engineering, civil engineering, applied mechanics, chemical engineering and aerospace engineering among others.

Constitutive modeling is a topic that is part of continuum mechanics and is broadly understood by engineers as a phrase that deals with the equations that describe the response of a material sample, when it is subjected to external loads. In recent times, the term *constitutive model* is used to describe any equation that attempts to describe a material response, either during deformation or failure, independent of its origin or mathematical structure.

In the research community, *constitutive modeling* and *continuum mechanics*, involve investigations of physical mechanisms in materials and mathematical frameworks to describe them. Eventual goal of these researches is to describe macroscopic response of materials. Simulators and designers, on the other hand, are interested in using constitutive models in their simulations. These investigations are more focused on obtaining quantitative estimates of material behaviour. Reasonableness of physical mechanisms, correctness of mathematical framework, simplicity of mathematical models, ease of numerical simulations and reliability of estimates are all important aspects of modelling of engineering materials. In view of these issues involved in modeling of materials, the authors felt that a compilation and presentation of a broad review is necessary for the use of students, researchers and practicing engineers. This is attempted in the current book. The book has the following special features:

- It introduces the basic principles of continuum mechanics, so that the user is familiar with the mathematical tools that are necessary to analyze finite deformations in materials. Special care is taken to ensure that there is an engineering flavour to the topics dealt with in continuum mechanics and only the mathematical details that are necessary to appreciate the physics and engineering of a given problem, are highlighted.

- A brief review of popular linear material models, which are used in engineering, and which are derived based on infinitesimal deformation of materials, is presented in the book.

- Popular material models that are used to characterize the finite deformation of solids and fluids are described.

- Some examples of continuum characterization of failure in solids, such as modeling using plasticity theory, degradation parameter etc. are presented in this book.

- Principles behind the constitutive modeling of few modern special materials such as shape memory materials and ferroelectric materials are presented using the basic principles of continuum mechanics.

- Detailed case studies are presented which include a complete description of the material, its observed mechanical behaviour, predictions from some popular models along with a detailed discussion on a particular model.

- A brief overview of the tools that are available to solve the boundary value problems, is also given in this book.

- Detailed exercise problems, which will help students to appreciate the applications of the principles discussed are provided at the end of chapters.

The book is an outcome of the teaching of a course called *Constitutive Modelling in Continuum Mechanics*, by the authors at IIT Madras. The graduate students taking this course consist of new material modelers as well as material behaviour analysts and simulators. Majority of them, however, are involved in selection and use of material models in analysis and simulation. Therefore, main goal of the course has been to expose students with various backgrounds to basic concepts as well as tools to understand constitutive models. While teaching this course, the authors experienced the need for a book where principles of continuum mechanics are presented in a simple manner and are linked to the popular constitutive models that are used for materials. Hence, lecture notes were written to meet the course objective and these lecture notes are now compiled in the form a book.

The authors were inspired by a continuous exposure to the latest issues in the field of continuum mechanics, which was made possible through efforts by a leading expert in the field of continuum mechanics; Prof. K. R. Rajagopal, Professor of Mechanical Engineering, Texas A&M University. Prof. Rajagopal, kept the flame of interest in continuum mechanics alive at IIT Madras, through his regular involvement in workshops, seminars and discussions. The authors deeply acknowledge the inspiration provided by him to the authors, as well as to many other students and faculty at IIT Madras.

The authors place on record the contributions made by many of their faculty colleagues in the shaping of this book. Prof. Srinivasn M. Sivakumar, Department of Applied Mechanics at IIT Madras, provided us with the notes on plasticity, which formed the basis for the discussions on plasticity that is presented in Chapter 6. The authors thank him for his valuable help. Prof. Raju Sethuraman, Department of Mechanical Engineering at IIT Madras, provided the basic ideas for the review of numerical procedures, and was a constant source of inspiration for the authors in completing this book. We deeply acknowledge his encouragement and support. We acknowledge the help of Profs. Sivakumar and Sethuraman, along with Dr. Mehrdad Massoudi, U.S. Department of Energy, in formulating the contents of this book.

This book would not have been made possible but for the willing contributions of a number of M. S., Ph.D. and M.Tech students, who were working with us during their stay in IIT Madras. We also acknowledge all the students of the course over years, because class discussions and class projects were helpful towards formulating contents as well presentation of the book.

The illustrations were drawn with great enthusiasm by Mr. Jineesh George and Mr. Santhosh Kumar. The work of Dr. Rohit Vijay during dual degree project, formed the basis of the case study on asphalt that is presented in Chapter 5. Ms. K. V. Sridhanya's MS thesis formed the basis for the case study on soils that is presented in Chapter 6. Dr. S. Sathianarayanan, who worked on piezo-polymers for his Ph.D. thesis, has helped us to put together the discussion on piezoelectricity in Chapter 7. Efforts of Mr. N. Ashok Kumar, Mr. D. Pandit and Mr. M. Kishore Kumar, who worked on shape memory matrials for their theses, have helped us in compiling the material in Chapter 7. Rajesh Nair has taken the pains of going through parts of manuscript and pointing out some errors. Mr. V. Srinivasan helped us in the cover design of the book. Mr. Jose Vinoo Ananth, Mr. Mohammed Ghouse, Mr. G.G. Uday

Kumar, Mr. Suresh Kumar have also contributed in various capacities in bringing out the final form of the book.

One of the authors (CLR) has utilized his sabbatical leave that is granted by IIT Madras, towards writing the first draft of the book. We greatly acknowledge the support offered by IIT Madras for the encouraging atmosphere that it offers to pursue scholastic ambitions like writing a book.

Our publishers Ane Books Inc., were patient enough to wait from the submission of our original proposal to publish this book. We greatly appreciate their encouragement and patience in finally bringing out the final form of this book.

Last but not the least, the authors acknowledge the time spared by their family members and other friends, directly or indirectly, for encouraging the authors to pursue this project.

C. Lakshmana Rao
Abhijit P. Deshpande

Notations

<u>**Symbols style**</u>

Regular, italicized	scalar variables, components and invariants of tensors, material constants
Boldface, small	vectors
Boldface, capital and Greek	tensors
Boldface, italics	vector or tensor material constants
$\equiv$	definition
$\wedge$	function
$*$	measurements made with reference to a moving frame of reference

<u>**Tensor operations**</u>

$\cdot$	dot product involving vectors and tensors
$\times$	cross product involving vectors and tensors
$\mathbf{ab}, \mathbf{vT}$	dyadic product of vectors $\mathbf{a}$ & $\mathbf{b}$, and vector $\mathbf{v}$ & tensor $\mathbf{T}$
$\mathbf{A}{:}\mathbf{B}$	scalar product of tensors $\mathbf{A}$ and $\mathbf{B}$ (double dot product)
$\mathbf{A}^{\mathrm{T}}$	transpose of $\mathbf{A}$
$|\mathbf{a}|$	magnitude of vector $\mathbf{a}$
$\det \mathbf{A}$	determinant of $\mathbf{A}$
$\mathrm{tr}(\mathbf{A})$	trace of $\mathbf{A}$
$\mathbf{A}^{-1}$	inverse of $\mathbf{A}$

<u>**Derivative operations**</u>

$\dot{s}(X,t),\ \dot{v}(X,t),\ \dot{T}(X,t)$	total (material or substantial) derivative with respect to time
$\dfrac{\partial s(x,t)}{\partial t},\ \dfrac{\partial v(x,t)}{\partial t},\ \dfrac{\partial T(x,t)}{\partial t}$	partial derivative with respect to time

$\overset{\circ}{\mathbf{T}}$	rotational derivative of $\mathbf{T}$
$\overset{\square}{\mathbf{T}}$	Jaumann derivative of $\mathbf{T}$
$\overset{\triangle}{\mathbf{T}}$	lower convected or covariant derivative of $\mathbf{T}$
$\overset{\triangledown}{\mathbf{T}}$	upper convected or contravariant derivative of $\mathbf{T}$
grad, div, curl	operators with respect to current configuration
Grad, Div, Curl	operators with respect to reference configuration
∇	gradient operator
∇^2	Laplace operator

List of symbols: Roman

A_x, A_X	areas in current and reference configuration, respectively
$\mathbf{a}$	acceleration
B^r	reference configuration
$\mathbf{b}$	body force
$\mathbf{b}^{em}$	electromechanical body force
$\mathbf{B}$	$\mathbf{V}^2$, left Cauchy Green tensor or Finger tensor
$\mathbf{B}_t$	$\mathbf{V}_t^2$
C_{ij}	material parameter associated with strain energy density function
$\mathbf{C}$	$\mathbf{U}^2$, right Cauchy Green tensor, matrix of elastic constants
C	stiffness coefficient
C^0	stiffness coefficient for biased piezoelectricity
$\mathbf{C}_t$	$\mathbf{U}_t^2$
$\mathbf{D}^v$	region (volume) in reference configuration
$\mathbf{C}^E$	electric current
D^r	Region (volume) in reference configruation
D^t	region (volume) in current configuration
$\mathbf{D}$	stretching tensor (rate of strain tensor, symmetric part of the velocity gradient tensor)
$\mathbf{D}^E$	electric displacement
e	strain
e^e, e^p	elastic and plastic strain
e_p	accumulated plastic strain, locked-in strain
$\dot{e}$	strain rate at small deformations
E	enthalpy, Young's modulus

E_r	relaxation modulus
E^*, E', E''	complex, storage and loss modulus
E_1, E_2	Burger's model parameters
$\mathbf{e}$	infinitesimal strain tensor
$\mathbf{e}^0$	biased infinitesimal strain tensor
$\mathbf{e}_i$	set of orthogonal unit base vectors,
$\dot{\mathbf{e}}$	strain rate tensor at small deformations
$\mathbf{E}$	Green strain, Electric field
$\mathbf{E}_t$	relative Green strain
$\mathbf{E}^{\mathbf{E}}$	electric field
$\mathbf{E}^{\mathbf{E0}}$	biased electric field
f	yield function
$\mathbf{f}_i$	set of orthogonal base vectors in a rotating frame
$\mathbf{f}_t$	force acting on region D^t
$\mathbf{F}$	deformation gradient
$\mathbf{F}^e$, $\mathbf{F}^p$	elastic and plastic deformation gradient
$\mathbf{F}_t$	relative deformation gradient
G	shear modulus, Doi model parameter
$\mathbf{g}$, g	acceleration due to gravity
g_{ij}, g^{ij}	metric coefficients
$\mathbf{g}_i$, $\mathbf{g}^i$	set of generalized base vectors
h	surface source of heat
$\dot{H}_t$	rate of heating
$\mathbf{H}$	displacement gradient
$\mathbf{H}_{\mathbf{L}}$	linear momentum
$\mathbf{H}_{\mathbf{A}}$	angular momentum
i, j, k	dummy indices
$I_{\mathbf{A}}, II_{\mathbf{A}}, III_{\mathbf{A}}$	first, second and third invariants of tensor $\mathbf{A}$, respectively
$\mathbf{I}$	unit tensor
J	Jacobian associated with $\mathbf{F}$
J_c	creep compliance
K	power law model parameter
$\mathbf{L}$	velocity gradient
M	degradation parameter
m_{D^t}	mass of the body in the sub-region D^t
$\mathbf{m}$	unit tangential vector
$\mathbf{m}^{\mathrm{em}}$	electromechanical body moment

M	total mass enclosed in a control volume D_t
$\mathbf{M}_t$	total moment
$\mathbf{n}$	unit normal vector
n	power law model parameter
N	number of cycles in cyclic plastic models
N_1, N_2	first and second normal stress difference
p	pressure, material particle
$p^{'}$	effective mean stress
$\mathbf{P}$	material polarization
q	effective deviatoric stress
q^i, q_i	set of generalised coordinates
Q	electric charge, state variable in plasticity
$\mathbf{q}$	heat flux vector
$\mathbf{Q}$	orthogonal tensor, state variables in plasticity
r	volumetric source of heat
R	radius of the yield surface in the octahedral plane
$\mathbf{R}$	rotation tensor
$\mathbf{R}_t$	relative rotation tensor
s	distance, length
s_v	kinetic variable
S^r	area in reference configuration
S^t	area in current configuration
$\mathbf{s}$	1st Piola Kirchhoff traction
$\mathbf{S}$	1st Piola Kirchhoff stress
$\mathbf{S}_1$	2nd Piola Kirchhoff stress
t	current time
t^r	time at which material body takes B^r
$t^{'}$	observation of time from a moving reference frame
$\mathbf{t}$	traction
u	pore pressure in soil mechanics
$\mathbf{u}$	displacement vector
$\mathbf{U}$	right stretch tensor
$\mathbf{U}_t$	relative right stretch tensor
V_x, V_X	volumes in current and reference configurations, respectively
v	volumetric strain
$\mathbf{v}$	velocity vector

$\mathbf{v}_p$	velocity of an object (projectile, particle etc.)
$\mathbf{V}$	left stretch tensor
$\mathbf{V}_t$	relative left stretch tensor
W	strain energy density
w_p	plastic work
$\dot{W}$	rate of work
$\dot{W}^{em}$	electromechanical rate of work
$\mathbf{W}$	spin tensor (vorticity tensor, skew-symmetric part of the velocity gradient)
$\mathbf{x}$	current configuration of a material point, representation of amaterial particle in real space
$\mathbf{x}'$	observation of the vector $\mathbf{X}$ from the moving reference frame
$\mathbf{x}^t$	configuration of a material point at time t
$\mathbf{X}$	reference configuration of a material point
y_i	set of orthogonal coordinates

List of symbols: Greek

α	scalar quantity
α_n	Ogden's material parameter
δ_{ij}	Kronecker delta
ε_{ijk}	alernator, alternating tensor
ε	internal energy
$\boldsymbol{\varepsilon}$	piezo electric coeficient matrix
$\boldsymbol{\varepsilon}^0$	piezo constant
$\in_0$	vacuum permittivity
ϕ	electric potential
γ	strain, shear strain
$\dot{\gamma}$	strain rate
η	entropy, stress ratio in soil mechanics
η_1, η_2	Burger's model parameters
κ	bulk modulus, mapping function between abstract and real configurations
κ_v	kinematic variable
κ	conductivity, dielectric constant
$\boldsymbol{\kappa}^0$	dielectric constant
λ	stretch or extension ratio, Lame's parameter, bulk or dilatational viscosity, plastic multiplier, structural parameter

λ_i	eigenvalues
μ	viscosity, Lame's parameter, Doi model parameter
μ_s	coefficient of static friction
$\overset{*}{\mu}, \mu', \mu''$	complex viscosity, real and imaginary parts of viscosity
μ_n	Ogden's material parameter
ν	Poisson's ratio
θ	temperature
θ_g	glass transition temperature
$\theta_l, \theta_h, \theta_{vh}$	low, high and very high temperatures to describe shape memory effect
ρ	density
σ_y	yield stress
$\boldsymbol{\sigma}$	Cauchy stress tensor
$\boldsymbol{\sigma}'$	effective stress in soil mechanics
τ	relaxation time, time, Doi model parameter
τ_{ret}	retardation time
$\boldsymbol{\tau}$	deviatoric stress tensor
ω	angular frequency
$\boldsymbol{\omega}$	infinitesimal rotation tensor
$\boldsymbol{\Omega}$	body spin tensor
ξ	internal variable
ψ	Helmholtz free energy
ζ	Gibbs free energy

List of symbols: Script

$\mathscr{B}$	body in abstract space
$\mathscr{E}$	Euclidean space
$\mathscr{p}$	material particle in abstract space
$\mathscr{R}$	real space

Contents

Introduction

आत्मा वा इदमेक एवाग्र आसीत्, नान्यत् किञ्चन मिषत् । (ऐतरेयोपनिषत्)
ātmā vā idameka evāgra āsit, nānyat kincana miṣat ... (Aitareyopaniṣad)

This existed as self-alone in the beginning. Nothing else winked.

1.1 INTRODUCTION TO MATERIAL MODELLING

All engineering materials are expected to meet certain performance requirements during their usage in engineering applications. These materials are often subjected to complex loadings, which could be in the form of a mechanical loading, a thermal loading, an electrical loading etc. or a combination of them. The response of the material to these loadings will determine the integrity of the material or the system in which the material is being used. A quantitative assessment of the material response when it is subjected to loads is very important in engineering design. This is possible if we have a mathematical description of the material response and its integrity, which can be called as a *material model*. The mathematical description of the system response, in the form of governing equations and boundary conditions, can be called as a *systems model*.

A model attempts to capture the underlying principles and mechanisms that govern a system behaviour through mathematical equations and is normally based on certain simplifying assumptions of the component behaviour. A model can typically be used to simulate the material as well as the system under different conditions, so as to predict their behaviour in situations where experimental observations are difficult. It is worth noting that in practice, we may have models that have a mathematical form without an understanding of physics, or models that describe the physics of the system, but may not be expressed in a specific mathematical form.

In what follows, we will outline the complexity of material and its response in engineering. Several modelling approaches, which attempt to understand and predict the material response, are also discussed briefly. In this overview, we will recollect many popular terms that are used in material modelling. These terms are italicized, without a definition at this stage. However, they will be defined more precisely in later chapters, along with concepts related to them.

1.2 COMPLEXITY OF MATERIAL RESPONSE IN ENGINEERING

Materials that are currently being used in engineering, are fairly complex in their composition as well as in their response. Following are few examples of such materials. Many engineering materials are *heterogeneous* in their composition, since they consist of different components or phases. For example, any concrete is truly a heterogeneous material with aggregates and a matrix material like a cement paste or asphalt. Materials exhibit different response when they are loaded and tested in different directions and hence are classified as *anisotropic*. Material composition can change through transformation *processes* such as chemical reaction and phase change. For example, a heterogeneous material may become homogeneous due to loading.

We will now outline few specific materials and their responses. Polymeric membranes, fiber reinforced composites are known to be anisotropic in their mechanical response. Many materials like polymers are '*viscoelastic*' in nature and exhibit a definite time dependent mechanical response. The same polymers show a time independent, large deformational response when they are deformed at temperatures above their '*glass transition temperature*'. We also know of the existence of special metals such as '*shape memory alloys*', which show drastic changes in their mechanical response when they are heated by about 50°C, causing a phase transition within the material. There are '*piezoelectric materials*' which are able to convert electrical energy to mechanical energy and *vice-versa*. Further, their electromechanical response is a function of the state of stress and the frequency of loading. Many engineering fluids show a '*linear stress-strain rate*' response, which is characterized by a parameter called as '*viscosity*'. However, there are other materials such as grease and paint, whose viscosity is dependent upon the state of stress at which the flow occurs. Blood clotting is a phenomenon where the material changes from a fluid to a solid. Mechanical response of blood during clotting can be understood only if biochemical reactions are also included in the model. The reasons for such complex material behaviour is also emphasized by analyzing *multiple time scales* of response and *multiple length scales* of response. The complexity of loadings, material make-up and its response is captured schematically in Fig. 1.1.

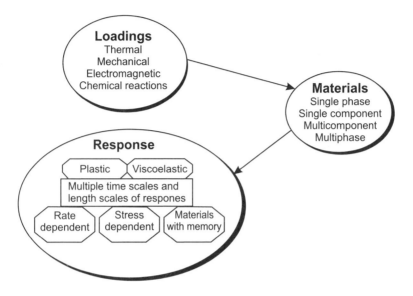

Fig. 1.1 Complexity of material response

It is always desirable to capture all the features that are observations of material response through a mathematical model. Clearly, a mathematical model for any material that can accurately capture the response observed in experiments for any of the materials listed above, is quite complex. The mathematical model that we operate with, should reflect our own understanding of the material response. For example, we know from history of strength of materials that earlier attempts were made to correlate the load applied on any solid to the elongation experienced by it. It took about hundred years of evolution to prove that this attempt is faulty and correlations should really be found between a concept called *stress* which is defined as load per unit area and a concept called *strain*, which is the deformation per unit length. A further evolution led to the visualization of stresses and strains as *second order tensors*. An assumption that these two tensors are linearly related, led to a formulation that is popularly known as *linear elasticity*. Experimental observations on materials like rubber, proved that *load measures* like stress and the *deformation measures* like strain will not always be related to each other linearly. The mechanical response of materials like rubber emphasized the need to introduce a *configurational (deformation)* dependence of stresses and the need for alternate deformational measures like *deformation gradients*. A redefinition of the *kinetic* (load related) measures and *kinematic* (deformation dependent) measures and their relationships are the main considerations in *continuum mechanics*. This framework is common to materials all classes of materials such as *solidlike*, *fluidlike* or gases.

It is worth noting that materials like metals and ceramics are clearly known to be *solids* and materials like water and oil are known to be *fluids*. Popularly, the response of solids has been considered through material model of linear elasticity. Similarly, the response of fluids has been considered through models of *Newtonian* or *inviscid fluid*. This in highlighted in Fig. 1.2. in the form of most widely used material models. On the other hand, polymers and granular materials are known to exhibit features of both solids and fluids. Hence, the use of terms such as solidlike and fluidlike is necessary to describe the response of materials.

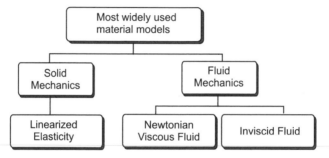

Fig. 1.2 Most widely used material models that are studied as part of solid mechanics and fluid mechanics

Mathematical framework for the description of the state of a material is formulated based on abstract notions and quantities. Abstract quantities such as force, velocity, stress and strain are used to define the state of a material. These quantities are visualized to be either scalars, vectors or tensors, having multiple components at any given point. However, experimental observations that can characterize the material, to the same detail as the mathematical framework, are generally not possible. For example, the only mechanical quantities that are measured for any material point are displacements and the time of observation. All other abstract quantities such as strains, velocities, accelerations, forces etc. are inferred from these basic observations. Consider, for example, experimental characterization of a piezo-electric material such as poly vinylidene flouride (PVDF). This material is primarily available and used in the form of thin sheets (25 – 100 mm). Testing of PVDF films in all the prescribed directions is not easy. Hence, often experimentalists perform some controlled experiments such as a uniaxial tension tests which provide data of load vs. longitudinal / transverse displacement. The constants that are demanded by a mathematical framework are often interpreted from the basic data collected from these simple tests. The interpretation of constants does lead a certain degree of uncertainty, since the interpretation of the same constant, for the same material, from two different tests may not always match with each other.

The mathematical models for any material can be assessed through comparisons with experimental observations. As mentioned above, these experimental observations are limited in nature. Hence, it is possible that there may be different mathematical models that are 'equally' successful in capturing the experimental observations. While it is necessary for a mathematical model to capture an experimentally observed phenomenon, this ability alone is not sufficient for the general applicability of the model in diverse situations. It is useful to classify different modelling approaches that are used in engineering practice. These are outlined in the next section.

1.3 CLASSIFICATION OF MODELLING OF MATERIAL RESPONSE

Before discussing different modelling approaches, let us first look at a specific material response and multiple ways of analyzing it. It is known that if a plastic (polymer) sample is deformed and kept at constant extension, the force required to maintain the extension decreases with time. Therefore, it is said that the *stress is relaxing* and the experiment is termed as a *stress relaxation* experiment. Now, one could look at the load *vs* time data taken from different materials and observe that decreasing load can be described by functional forms such as exponential or parabolic. In this case, no hypothesis is made about the material behaviour and no detailed justification is given about why a particular functional form is chosen. The constants used in the functions will be different for different materials and can therefore be used to distinguish material behaviour. We will call such approaches to modelling of materials as empirical modelling.

Let us continue with our example and compare the response of the polymer in stress relaxation with other well known materials, such as steel or water. An observation can be made that the polymer response is in some way a *combination of the responses* of these two types of responses, namely elastic and viscous. Therefore, one can make hypothesis about material being viscoelastic and construct mathematical model, which in certain limits reduces to elastic or viscous behaviour. Such models will be called phenomenological models, because the overall material response serves as a guide in building of the models. An example of such model is Maxwell model, which predicts that stress will decrease exponentially in a stress relaxation experiment. The constants used in the exponential form can be called *material constants* of Maxwell model, as they will be different for different materials.

With increasing theoretical development at the microscopic scale and computational resources, we can talk of another set of models, *i.e.*, *micromechanical models*. Such a model draws recourse to the make-up of material in its more elementary forms such as atoms, molecules, agglomerates, networks, phases etc. In our example of stress relaxation in a plastic, polymer would be considered as a collection of molecular segments. A hypothesis can be made about the mechanical response of a segment. The response of bulk polymer can be obtained if we are able to develop a mathematical model for a collection of polymer segments. Of course, such a model will also lead to decreasing stress at the bulk scale and material constants at the bulk scale.

More often than not, it is a combination of these approaches, empirical, phenomenological and microscopic, that is used by engineers to understand and predict material behaviour. Each of them is useful in a specific context. In the following discussion, we outline their strengths and limitations.

1.3.1 Empirical Models

In engineering, many of the procedures and practices are also dictated by documents called *design codes and standards*. These documents are normally a compendium of human experience, documented for use by a practitioner with the least difficulty. In development of such documents, all uncertainties and ambiguities in human experience, are also accounted for, so as to help to develop a *safe* design. Since the design codes are meant to be used by a common practitioner, they must necessarily use concepts that are more easily grasped by a common practitioner. The use of multiple components of stress tensor in all practical situations is difficult for a practitioner and hence, the three dimensional nature of stress is often captured in a convenient scalar stress measure such as an *equivalent stress*. Similarly, *equivalent uniaxial strain* measures are defined and sometimes a relation is sought between these defined equivalent measures. Even though these relationships may not have strict mathematical validity, they are useful in characterizing a material, especially when we want to characterize the material response due to complex time dependent loading conditions. We could call these equations as *empirical models*. The empirical models, by and large are curve fits of available experimental data. They will be very useful in design and are applicable within the range of data from which they have been derived. However, they have no basis in either the physics of deformation of the material, or in the mathematical rigor or accuracy of the variables that they are attempting to correlate. Such approaches are also adopted by researchers when they are handling new materials, whose response is not yet fully understood, and to obtain quick approximate description of material behaviour.

In recent times, an approach based on Artificial Neural Network (ANN) is being used to describe the material behaviour. A class of artificial neural networks, known as MLFFNN (Multilayer Feed Forward Neural Network) is being used to correlate the microstructural parameters with macroscopic mechanical behaviour. This ability of MLFFNNs is attributed to the presence of non-linear response units and the ability of the network to generalise from given number of experimental observations. This is the primary motivation behind the choice of ANNs for prediction of material response. In the use of ANN for material modelling, the material behaviour is no longer represented mathematically but is described by neuronal modelling. The main aim of ANN modelling, is to build a neural network directly from experimental results. The prediction accuracy of such models is largely dependent on the training schemes that are used in the building of the ANN model.

1.3.2 Micromechanical Models

Attempts have been made to construct models of material's *macroscopic response* from the mechanics of deformation of their individual units. Here, we define a macro response as the response that is observed on units where mechanical measures such as stresses and strains are defined and are valid. Normally these mechanical quantities are defined and are valid over *representative volume elements*. The size of the element is such that it is not necessary to consider the finer details of the material constitution, while working with the measures of stresses and strains that are defined at the level of a representative volume element. As an example, we could visualize the characterization of stress field in a composite material like cement concrete. Concrete is a material that consists of cement paste, course aggregates and fine aggregates. If we want to treat concrete as a single material, by ignoring the finer details of its constituents, we must define and use stress and strain measures at scales that are larger than the size of aggregates, to interpret the results obtained from such models. Hence, in a concrete material having an aggregate with an average size of 3 mm, stresses, strains and their relationships would be valid at scales of about 30 mm.

If we attempt to describe the behaviour of a material at a scale that is below the scale of a representative volume element, we need to recognize its constituent parts and their possible interactions and postulate as to how, these interactions would result in a phenomenon observed at the level of the representative volume element. Any attempt that tries to deal with a material at a level that is finer than the scale of the representative volume element, we will classify as a micromechanical modelling of the material. The size of representative volume elements, their constituents and their scales for typical engineering materials are listed in Table 1.1.

A distinct advantage of a micromechanical model is that it recognizes the various constituents and the mechanics of their interactions, which may form the basis of an observed macro-phenomenon. However, it is not always possible to find a direct correlation between an observed macro response and a postulated micromechanical interaction mechanism. In other words, there may be several micromechanisms, which will manifest in a single macro observation. Further, a micromechanism that is able to correlate with a single macro-phenomenon, may not explain a different macro-phenomenon. For example, statistical network model of chain motion was used to explain the shear stiffness observed in rubber elasticity. However, the same model was not found to be sufficient to characterize the tensile and visco-elastic behaviour of the same material.

Table 1.1 Size of the Representative Volume Element

Representative	Volume Element
Metals and ceramics	$(0.1 \text{ mm})^3$
Polymers and most composites	$(1 \text{ mm})^3$
Wood	$(1 \text{ mm})^3$
Concrete	$(100 \text{ mm})^3$

1.3.3 Phenomenological Models

The phenomenological models of material response do not consider the details of the material structure explicitly, but postulate the material response to be the manifestations within a defined mathematical framework, that is valid for a class of materials. These mathematical frameworks will be in the form of mathematical relations having undetermined parameters. These parameters are normally correlated with experimental observations. The parameters themselves may or may not have any physical significance, since the relations are not always formulated by considering the underlying mechanisms.

However, the phenomenological models are more universal in their appeal since they can be used directly in numerical simulations in multidimensions. The phenomenological models are also known as the constitutive models and are used in conjunction with governing equations in continuum mechanics, which are formulated at the scale of representative volume elements. Some of these models attempt to capture the micromechanical details associated with the deformation of materials, so that the individual characteristics of specific materials are not lost while capturing the overall material response.

In many solidlike materials, it is important to capture not only the deformational response of the material, but also to describe the failure of materials which is in the form of material separation, or very large deformations. Failure of solids, is normally due to the accumulation of many defects, which may form a crack and the unstable propagation of this crack leads to material separation. The deterioration of a material within a phenomenological framework is attempted by treating damage as a continuous variable and by monitoring the progress of this variable as the material deforms. Attempts to characterize failure in solids through conditions for unstable crack propagation are also attempted within the field of fracture mechanics.

Many classes of material responses are modeled using phenomenological models. Any given material may exhibit different classes of material response, depending on the conditions under which the material is being used. A summary of the common materials used in engineering and their common material response is given in Table 1.2. It may be seen in the table that many common materials like metals exhibit an elastic response when they are loaded below their plastic limit, and will exhibit a fluid like visco-plastic behaviour when they are loaded beyond their plastic limit. PVDF exhibits rubber like behaviour of a polymeric material as well as an elctromechanical response.

Table 1.2 Common Materials and their Class of Material Response

Materials	Class of Material Response
Metals	Linear elastic, plasticity, viscoplastic
Water, Oil	Newtonian
Air	Viscous, inviscid
Concrete	Linear elastic, plasticity
Soils	Hyperelastic, hypoelastic
Polymers	Linear elastic, viscoelastic, plasticity
Rubbers	Hyperelastic
Blood	Non-Newtonian
Asphalt	Viscoelastic, non-Newtonian
Ferroelectrics	Electromechanical
Shape memory alloys	Thermomechanical
Paints, emulsions, foods	Non-Newtonian, viscoelastic, thixotropic

In engineering practice, the phenomenological models are used along with the governing equations and appropriate boundary conditions, so as to be in a position to yield a system response that will be useful in an application. The solution of the system of equations, is obtained by using approximate methods like finite element methods. Many times, the use of these approximations will yield a set of non-linear algebraic equations that may need to be solved iteratively, in order to yield a reasonable system response. The engineer normally compares these predicted responses with experimental observations made at a system level, in order to assess the validity of his simulations using phenomenological models.

1.4 LIMITATIONS OF THE CONTINUUM HYPOTHESIS

The fundamental assumption of the continuum approximation is that quantities vary slowly over lengths on the order of the atomic scales (10^{-6}–10^{-8} m).

Concepts like stresses or strains are meaningful only in a length and time scales where variations are relatively slow and the number of participating atoms are of the order of 10^6–10^8. Hence, when we talk of materials having a microstructure of nano-scale, which undergo deformations of the order of nanometers, The continuum assumption of the material may no longer be valid. At such microscales, it is useful to simulate the material deformation using *molecular dynamics (MD) simulation*. MD simulation is essentially a particle method where the governing equations of particles based on Newton's laws are written down for a group of atoms that are assumed to constitute the material. In fact, simulation of fracture processes in many materials, near the crack tip is currently being attempted using this approach. At even lower length scales, the visualization of matter as atoms having a certain mass also fails. In this visualization, the state of a particle gets defined by a wave function based on a concept of *wave-particle duality*. The governing equations at this scale is the Schrodinger equation and solutions for this equation are attempted by many researchers working in the field of quantum mechanics.

1.5 FOCUS OF THIS BOOK

This book is based on the premise that the primary goal of engineers while dealing with material response is to simulate and predict material behaviour in real situations. One advantage of simulations is that they do not have limitations of scale. However, they are limited by the extent to which physical phenomena are captured in the mathematical framework and by the capacity of computational tools. Traditionally, designers relied on full-scale experiments wherever they were possible. However, for the systems that are more complicated, the experimental tools by themselves are not sufficient to ensure a successful design. Hence, simulations are necessary. Simulations are not possible without mathematical model. An important component of the mathematical modelling is to model the material behaviour itself.

In this book, we will attempt to give a bird's eye view of material models from a phenomenological standpoint. Therefore, the emphasis will be to provide an overview of phenomenological models defined above. One of the reasons for this standpoint is that simulations in multidimensions are most often carried out using a continuum framework with phenomenological models. Some of these phenomenological models are inspired or are derived from micromechanical models. It is also noteworthy that empirical models are the precursors to the formulation of phenomenological models. Therefore, empirical and micromechanical models will also be examined in this book, wherever necessary.

The preliminary mathematical concepts required for the discussion of material modelling are reviewed in Chapter 2. We will review the basic framework of continuum mechanics and the place of constitutive equations within this framework in Chapter 3. Emphasis in this chapter will be on the description of general concepts which are valid for all materials. Specific constitutive models that are popularly used in fluids and solids will be explained in details over the next few chapters. In Chapter 4, we will present the *simplistic models* for different class of material behaviour. These classes include *elastic*, *viscous*, *viscoelastic* and *elastoplastic* materials. An overview of the *non-linear models* of fluids is the subject of Chapter 5. Chapter 6 focuses on models to describe the *non-linear* behaviour of solids. Thermomechanical and electromechanical response of engineering materials is discussed in Chapter 7. Finally, a brief review of solution methods of the governing equations is presented in Chapter 8, along with concluding remarks.

The focus of the book is primarily on introducing all aspects that are involved in the complex topic of material modelling, rather than going into the intricate details of any one of them. A bibliography is provided for any reader who is interested in more details of any one of the aspects covered in the book.

■■■

2

Preliminary Concepts

... सत्यम् च स्वाध्याय प्रवचने च। तपश्च स्वाध्याय प्रवचने च।
दमश्च स्वाध्याय प्रवचने च। शमश्च स्वाध्याय प्रवचने च। .... *(तैत्तिरीयोपनिषत्)*

*... satyam ca svādhyāya pravachane ca, tapaśca svādhyāya
pravacane ca,*
 *damaśca svādhyāya pravachane ca, śamaśca svādhyāya pravacane
ca ....* *(Taittirriyopanisad)*

...truth also is to be practiced and taught, contemplation also is to be
practiced and taught, external restraint also is to be practiced and taught,
internal restraint also is to be practiced and taught ...

2.1 INTRODUCTION

Modelling of engineering materials involves analysis with physical variables,
such as mass density, displacement, velocity, force, stress, strain, magnetic field
and current density. In fact, analysis is nothing but finding and understanding
relations among these variables. As mentioned in the Section 1.1, the system
model consists of governing equations and boundary conditions. Governing
equations are written for field variables such as temperature, velocity and stress.
Most of us are familiar with temperature as a scalar (magnitude but no direction),
velocity as a vector (magnitude and one direction) stress as a tensor (magnitude
and two directions). Scalars, vectors and tensors as described above are examples
of mathematical objects called tensors, which are described in more detail below.

2.2 COORDINATE FRAME AND SYSTEM

Physical variables described above are defined for a material at different points.
We need a system to describe all the points within the material to be able to
describe the physical laws among the physical variables. Frame and coordinate

system are used to describe the material in space. A frame implies an observer, therefore implying only the origin of observation. A coordinate system, on the other hand, implies an origin and geometric specification of all the points in space relative to that origin. Therefore, it is possible to have multiple frames being described in one coordinate system. Also, it is possible to describe observations from a single frame using many alternate coordinate systems.

It is natural for us to expect that it should be possible to describe physical laws and material behaviour in different frames and different coordinate systems. However, we would also expect that physical laws and material behaviour are not affected by how we observe or describe them. This independence from frame and coordinate description is termed as *invariance* of physical laws or material behaviour. Several definitions and requirements of invariance are possible and they will be discussed in more detail as and when needed. In summary, we should note that in modelling of engineering materials, it is of interest to talk both of *frame invariance* (*also called frame indifference*) or *invariance with respect to coordinate transformations*.

Rectangular coordinate system is used often to describe physical variables and physical laws. In many engineering applications, *curvilinear coordinates* such as cylindrical and spherical coordinate systems are more convenient and are therefore, utilized. All of these are examples of orthogonal coordinate system. Given the discussion in the above paragraph, many other coordinate systems are used to describe governing equations in modelling of engineering materials. *Generalized coordinate system* implies a completely general set of coordinates and *base vectors*. The *converced frame* and *convected coordinates* are also used to describe the material behaviour. These are described in detail in Appendix.

2.3 TENSORS

Tensor is derived from *tensio* (Latin for tension, stress). **T**, a tensor of order n has components:

$$T_{ijk} \dotfill \text{upto } n, \qquad\qquad i = 1, 2, \dotfill d, \qquad\qquad (2.3.1)$$

where d is the number of dimensions. In other words, for a tensor of order n, the number of subscript indices is n, where $n = 0, 1, 2, 3, \dots$ For example, tensor of order 0 is a scalar and tensor of order 1 is a vector.

Tensors have property that it is always possible to transform the components of a tensor to another coordinate system. Let us denote the components thus transformed as follows:

$$T'_{rst} \dotfill \text{upto } n \qquad\qquad i = 1, 2, \dotfill d. \qquad\qquad (2.3.2)$$

Tensors of order 0, 1 and 2 are most frequently encountered in modelling of engineering materials. We will restrict our attention to three dimensional physical space, so the indices $i, j, k....$ would take values of 1, 2 and 3. In the following discussion, T and T' refer to the functions in two different coordinate systems, respectively.

2.3.1 Tensors of Different Orders

It is worth noting that tensors of the lower orders as defined below have special meaning in engineering analysis. We will therefore take a closer look at these tensors below.

Order 0

A tensor of order 0 would imply that there are no subscripts to define the tensor. This would mean that this tensor has only a magnitude. The observation of this tensor from any other co-ordinate system would not change the tensor. Conventionally, we know that such quantities are called scalars and they are invariant with respect to any co-ordinate transformation. A few of the common examples of zeroth order tensors are density, temperature, length, specific heat, energy and electrical charge.

Order 1

A tensor of order 1 would imply one subscript. Hence the components of this tensors would be denoted by

T_i, where $\quad i = 1, 2$ and 3. $\hspace{4cm}$ (2.3.3)

The tensor itself is denoted in bold face as **T**, while the components are denoted without the bold face. The first order tensors are conventionally known as vectors. We are familiar with several examples of vectors such as velocity, force, heat flux, electrical field etc. A vector **T** has three components T_1, T_2 and T_3. In another coordinate system, the same vector can be denoted as **T'** and will have three components as T_1', T_2', and T_3'.

We define the magnitude of a vector **T** as

$$|\mathbf{T}| = \sqrt{T_1^2 + T_2^2 + T_3^2} \, .$$ (2.3.4)

Similarly, the magnitude of **T'** will be

$$|\mathbf{T'}| = \sqrt{T_1'^2 + T_2'^2 + T_3'^2} \, .$$ (2.3.5)

A property of the first order tensors is that

$$|\mathbf{T}| = |\mathbf{T'}|.$$ (2.3.6)

In other words, the magnitude of a tensor of order one is invariant with respect to coordinate transformation.

Order 2

Tensor of order two, will have two subscripts. Hence, the components of this tensor will be denoted as T_{ij}. The same tensor, when denoted with respect to a different coordinate system will be denoted as T'_{ij}.

$$\mathbf{T} = \begin{pmatrix} T_{11} & T_{12} & T_{13} \\ T_{21} & T_{22} & T_{23} \\ T_{31} & T_{32} & T_{33} \end{pmatrix}. \tag{2.3.7}$$

Stress, strain and polarisability are examples of second order symmetric tensors. In another coordinate system, the tensor would be denoted by $\mathbf{T'}$ and the components as T'_{ij}.

Many second order tensors that we use in engineering practice, have the following property:

$$T_{ij} = T_{ji}. \tag{2.3.8}$$

Tensors having the above property are called as *symmetric* tensors. Some examples of symmetric tensors that we will encounter later are *stress*, *strain*, *stretching* or s*train rate tensor* etc. Second order symmetric tensors will have only *six* independent components as illustrated below:

$$\mathbf{T} = \begin{pmatrix} T_{11} & T_{12} & T_{13} \\ T_{12} & T_{22} & T_{23} \\ T_{13} & T_{23} & T_{33} \end{pmatrix}. \tag{2.3.9}$$

It can be proved that if T_{ij} of the symmetric tensors defined above are real, there will exist three scalars that are defined as eigenvalues of $\mathbf{T}$. The three eigenvalues are real numbers and are denoted by λ_1, λ_2 and λ_3. These can be evaluated from a *characteristic equation* that is given below:

$$|\mathbf{T} - \lambda\mathbf{I}| = 0 \tag{2.3.10}$$

The eigenvalues of a tensor remain the same even if the tensor is transformed to another coordinate system. Therefore, for the tensor $\mathbf{T}$, λ_1, λ_2 and λ_3 are invariants and any combination of these three is also invariant. Three invariants of a tensor $\mathbf{T}$ that are defined in terms of the eigenvalues are given below: (2.3.11).

$$\left.\begin{aligned} I_T &= \lambda_1 + \lambda_2 + \lambda_3 \\ II_T &= \lambda_1\lambda_2 + \lambda_2\lambda_3 + \lambda_3\lambda_1 \\ III_T &= \lambda_1\lambda_2\lambda_3 \end{aligned}\right\} \tag{2.3.11}$$

The above invariants can also be written in terms of the tensor components T_{ij}. Such descriptions are very commonly used in engineering practice. These definitions are given below:

$$\left.\begin{array}{l} I_T = \sum\limits_{i} T_{ii} \\[2mm] II_T = \sum\limits_{i,j} T_{ij} T_{ji} \\[2mm] III_T = \sum\limits_{i,j,k} T_{ij} T_{jk} T_{ki} \end{array}\right\} \qquad (2.3.12)$$

The characteristic equation can also be understood based on transformation of a vector by $\mathbf{T}$. The transformation of an arbitrary vector $\mathbf{v}$ by $\mathbf{T}$ is given by $\mathbf{T} \cdot \mathbf{v}$ (The details of this transformation is discussed later in this chapter). If this transformation results only in stretching, then the transformed vector is also just a scalar times the original vector, *i.e.*, $\lambda \mathbf{v}$. Therefore, eigenvalues of $\mathbf{T}$ are understood in terms of their transformations that are affected by an arbitrary vector which is also called as *eigenvector*. This transformation is mathematically expressed below:

$$\mathbf{T} \cdot \mathbf{v} = \lambda \mathbf{v} \quad \text{or} \quad \mathbf{T} \cdot \mathbf{v} - \lambda \mathbf{v} = 0. \qquad (2.3.13)$$

Cayley-Hamilton theorem states that a tensor satisfies its own characteristic equation:

$$\mathbf{T}^3 - I_T \mathbf{T}^2 + II_T \mathbf{T} - III_T \mathbf{I} = \mathbf{0} \qquad (2.3.14)$$

This implies that $\mathbf{T}^3$ can be expressed as a function of $\mathbf{T}^2$ and $\mathbf{T}$ ($\mathbf{T}^3 = \mathbf{T} \cdot \mathbf{T} \cdot \mathbf{T}$ and $\mathbf{T}^2 = \mathbf{T} \cdot \mathbf{T}$, operation described in Table 2.2). In fact, any higher order power of $\mathbf{T}$ can be expressed in terms of $\mathbf{T}^2$ and $\mathbf{T}$. In later chapters, we will make extensive use of this property of tensors.

The next section gives a summary of operations with vectors and tensors and notation used for them.

2.3.2 Notations for Tensors

This section is intended to describe the rules of *index notation* or *suffix notation*, which will be used while describing mathematical relations. We are familiar with the three unit vectors used in rectangular co-ordinate system, and are popularly denoted as $\mathbf{i}$, $\mathbf{j}$ and $\mathbf{k}$. These vectors are also known as *base vectors* and are denoted as $\mathbf{e}_x$, $\mathbf{e}_y$ and $\mathbf{e}_z$ or $\mathbf{e}_1$, $\mathbf{e}_2$ and $\mathbf{e}_3$. In short, the base vectors are also denoted as follows:

$$\mathbf{e}_i, \quad \text{where } i = 1, 2, 3. \qquad (2.3.15)$$

We know that vectors such as **a** and **b** can be represented in terms of the base vectors as follows:

$$\mathbf{a} = \sum_{i=1}^{3} a_i \mathbf{e}_i = a_1 \mathbf{e}_1 + a_2 \mathbf{e}_2 + a_3 \mathbf{e}_3$$

$$\mathbf{b} = \sum_{j=1}^{3} b_j \mathbf{e}_j \qquad (2.3.16)$$

Here, i and j are called *dummy indices* which take values of 1, 2 and 3. For an orthogonal coordinate system:

$$\mathbf{e}_1 \cdot \mathbf{e}_1 = 1$$
$$\mathbf{e}_1 \cdot \mathbf{e}_2 = 0 \qquad (2.3.17)$$

It should be recalled that we have used the familiar dot product of vectors (also called the scalar product) in the above equation.

In general, we can write that

$$\mathbf{e}_i \cdot \mathbf{e}_j = 0 \quad \text{for } i \neq j$$
$$= 1 \quad \text{for } i = j \qquad (2.3.18\ (a)$$

This result can be written in terms of Kronecker delta δ_{ij}, which is defined as follows:

$$\delta_{ij} = 0 \quad \text{for } i \neq j$$
$$= 1 \quad \text{for } i = j \qquad (2.3.18b)$$

Therefore, $\qquad\qquad \delta_{ij} = \mathbf{e}_i \cdot \mathbf{e}_j \qquad\qquad (2.3.19)$

Similarly, cross products of orthogonal base vectors can be represented by ε_{ijk} (called *alternator* or *alternating tensor*, though it is not a tensor),

$$\mathbf{e}_i \times \mathbf{e}_j = \varepsilon_{ijk} \mathbf{e}_k \qquad (2.3.20)$$

ε_{ijk} takes values depending on values of i, j and k :

$$\varepsilon_{ijk} = 1 \quad \text{if } ijk = 123,\ 231,\ 312$$
$$= -1 \quad \text{if } ijk = 321,\ 132,\ 213 \qquad (2.3.21)$$
$$= 0 \text{ if any two indices are equal.}$$

Therefore, we can write the cross product of two vectors using ε_{ijk}

$$\mathbf{a} \times \mathbf{b} = \sum_{i} a_i \mathbf{e}_i \times \sum_{j} b_j \mathbf{e}_j$$

$$= \sum_{i,j,k} a_i b_j \varepsilon_{ijk} \mathbf{e}_k \qquad (2.3.22)$$

Example 2.3.1: Show that $\mathbf{a} \cdot \mathbf{b}$ can be represented as $a_i b_i$ in index notation. Expand the result given in index notation.

Solution:

As given in Eqn. (2.3.16) writing $\boldsymbol{\sigma}$ and $\mathbf{T}$ in index notation:

$$\mathbf{a}\cdot\mathbf{b} = \sum_i a_i e_i \cdot \sum_j b_j e_j$$

$$= \sum_{i,j} a_i b_j \left(e_i \cdot e_j\right)$$

The dot product operation is carried out between adjoining neighbours of the dot product operator (e_i and e_j). Therefore, we obtain from Eqn. (2.3.19),

$$\mathbf{a}\cdot\mathbf{b} = \sum_{i,j} a_i b_j \delta_{ij}$$

Using properties of the Kronecker delta from Eqn. (2.3.18),

$$\mathbf{a}\cdot\mathbf{b} = \sum_i a_i b_i = \sum_j a_j b_j$$

Expanding both the sum, we get the result

$$\mathbf{a}\cdot\mathbf{b} = a_1 b_1 + a_2 b_2 + a_3 b_3$$

Following are couple of useful identities relating ε_{ijk} and δ_{ij}:

$$\sum_j \sum_k \varepsilon_{ijk}\varepsilon_{hjk} = 2\delta_{ih}$$

$$\sum_k \varepsilon_{ijk}\varepsilon_{mnk} = \delta_{im}\delta_{jn} - \delta_{in}\delta_{jm} \qquad (2.3.23)$$

In index notation, $\mathbf{a} = \sum_{i=1}^{3} a_i e_i$ is written as $a_i e_i$ by omitting the summation. This is due to the convention that whenever an index is repeated, a summation is implied. An equation can be written in terms of vectors or tensors themselves or in terms of their components. Equations written in the former format will be referred to as the equations written in boldface notation, and equations written with the later format will be referred to as equations written in index notation or suffix notation. The suffix notation is used for compactness and ease of manipulating relations. Using these conventions, some of the operations are summarized in Table 2.1.

Table 2.1 Operations with Vectors and their Notation

Operation	Boldface Notation	Index Notation
Multiplication of vector and a scalar	$s\mathbf{a}$	sa_i
Scalar product of two vectors (also called dot product)	$\mathbf{a} \cdot \mathbf{b}$	$a_i b_i \delta_{ij} = a_i b_i = a_j b_j$
Cross product of two vectors	$\mathbf{a} \times \mathbf{b}$	$a_i b_j \varepsilon_{ijk}$
Determinant of $\mathbf{A}$	det $\mathbf{A}$	$\varepsilon_{ijk} A_{1i} A_{2j} A_{3k}$

As shown in Eqn. (2.3.7), nine components of the tensor can be represented as a 3×3 matrix. A tensor is represented in index notation as

$$\mathbf{T} = T_{ij} \mathbf{e}_i \mathbf{e}_j \tag{2.3.24}$$

The unit tensor is defined as

$$\mathbf{I} = \delta_{ij} \mathbf{e}_i \mathbf{e}_j = \begin{bmatrix} 1 & 0 & 0 \\ 0 & 1 & 0 \\ 0 & 0 & 1 \end{bmatrix} \tag{2.3.25}$$

Some of the operations with tensors of different orders are shown in Table 2.2.

Table 2.2 Operations with Tensors and their Notation

Operation	Boldface Notation	Index Notation
Dyadic product of two vectors $\mathbf{a}$ and $\mathbf{b}$	$\mathbf{ab}$	$a_i b_j$
Multiplication of a scalar to a tensor	$s\mathbf{T}$	sT_{ij}
Addition of two tensors	$p\mathbf{I} + \boldsymbol{\sigma}$	$p\delta_{ij} + \sigma_{ij}$
Scalar product of tensors (double dot product)	$\boldsymbol{\sigma} : \mathbf{T}$	$\sigma_{ij} T_{ji}$
Dot product of a tensor and vector	$\mathbf{T} \cdot \mathbf{v}$	$T_{ij} v_j$
Dot product of two tensors	$\boldsymbol{\sigma} \cdot \mathbf{T}$	$\sigma_{ij} T_{jm}$

Example 2.3.2: Show that $\sigma : \mathbf{T}$ can be represented as $\sigma_{ij} T_{ji}$ in index notation. Expand the result in index notation and write the complete result.

Solution:

Double dot product or scalar product of two tensors implies two dot product operations (see *Example* 2.3.1). As given in Eqn. (2.3.24) writing σ and $\mathbf{T}$ in index notation:

$$\sigma : \mathbf{T} = \sigma_{ij} \mathbf{e}_i \mathbf{e}_j : T_{lm} \mathbf{e}_l \mathbf{e}_m$$

$$= \sigma_{ij} T_{lm} (\mathbf{e}_i \mathbf{e}_j : \mathbf{e}_l \mathbf{e}_m)$$

The first dot product operation is carried out between adjoining neighbours of the double dot product operator ($\mathbf{e}_j$ and $\mathbf{e}_l$). Therefore, we obtain

$$\sigma : \mathbf{T} = \sigma_{ij} T_{lm} \delta_{jl} (\mathbf{e}_i \cdot \mathbf{e}_m)$$

Using properties of the Kronecker delta Eqn. (2.3.18),

$$\sigma : \mathbf{T} = \sigma_{ij} T_{jm} (\mathbf{e}_i \cdot \mathbf{e}_m)$$

Now, the second operation can be carried out and we get

$$\sigma : \mathbf{T} = \sigma_{ij} T_{jm} \delta_{im}$$

Again, using the properties of Kronecker delta, we get

$$\sigma : \mathbf{T} = \sigma_{ij} T_{ji}$$

$$\sigma : \mathbf{T} = \sigma_{11} T_{11} + \sigma_{12} T_{21} + \sigma_{13} T_{31} + \sigma_{21} T_{12} + \sigma_{22} T_{22} +$$

$$\sigma_{23} T_{32} + \sigma_{31} T_{13} + \sigma_{32} T_{23} + \sigma_{33} T_{33}$$

Example 2.3.3: Show that a dot product of a tensor and a vector as described in Table 2.2 is also a vector. Therefore, this dot product could be understood as an operation in which *a tensor transforms an arbitrary vector into another vector*.

Solution:

Dot product operation of a tensor and a vector is given by,

$$\sigma : \mathbf{T} = T_{mn} \mathbf{e}_m \mathbf{e}_n \cdot a_p \mathbf{e}_p$$

$$= T_{mn} a_p \mathbf{e}_m (\mathbf{e}_n \cdot \mathbf{e}_p)$$

As mentioned earlier, the operation is carried out between adjoining neighbours of the operator. Therefore, we obtain

$$\mathbf{T} \cdot \mathbf{a} = T_{mn} a_p \delta_{np} \mathbf{e}_m = T_{mp} a_p \mathbf{e}_m$$

The resulting quantity is a variable, which can be written in component form as

$$T_{mp} a_p \mathbf{e}_m = \left(T_{11} a_1 + T_{12} a_2 + T_{13} a_3 \right) \mathbf{e}_1 + \left(T_{21} a_1 + T_{22} a_2 + T_{23} a_3 \right) \mathbf{e}_2 + $$
$$\left(T_{31} a_1 + T_{32} a_2 + T_{33} a_3 \right) \mathbf{e}_3$$

If we refer to the transformed vector as $\mathbf{b}$, then

$$b_1 = \left(T_{11} a_1 + T_{12} a_2 + T_{13} a_3 \right)$$
$$b_2 = \left(T_{21} a_1 + T_{22} a_2 + T_{23} a_3 \right)$$
$$b_3 = \left(T_{31} a_1 + T_{32} a_2 + T_{33} a_3 \right)$$

Reminding ourselves about the vector $\mathbf{a}$ in component form,

$$\mathbf{a} = a_1 \mathbf{e}_1 + a_2 \mathbf{e}_2 + a_3 \mathbf{e}_3$$

Therefore, we can observe that the effect of the operation is to transform the vector $\mathbf{a}$ to another vector.

2.4 DERIVATIVE OPERATORS

Several derivative operators are used in modelling of engineering materials. This section summarizes them and their description in index notation. Conventionally, index notation is used for rectangular coordinates. This is being developed in detail in this section. The equivalent quantities in curvilinear coordinates are given in Appendix.

We know that for rectangular Cartesian co-ordinates, the gradient operator is given as:

$$\text{grad} \equiv \mathbf{e}_1 \frac{\partial}{\partial x_1} + \mathbf{e}_2 \frac{\partial}{\partial x_2} + \mathbf{e}_3 \frac{\partial}{\partial x_3} \tag{2.4.1}$$

In certain sources, ∇ is used instead of grad to indicate the gradient operator. In the above equation, $\equiv$ implies an equivalence and is used to define quantities.

A field is a physical variable which is a function of space and time. In modelling of engineering materials, several operations of the gradient operator on tensors are used. These operations are described in Table 2.3. The operations and the detailed components are given in Table 2.4 for cylindrical coordinates and in Table 2.5 for spherical coordinates.

Table 2.3 Operations with Gradient Operator

Operation	Boldface notation	Index notation
Gradient of a scalar field	grad s	$\dfrac{\partial s}{\partial x_i}$
Divergence of a vector field	div $\mathbf{v}$	$\dfrac{\partial v_i}{\partial x_i}$
Curl of a vector field	curl $\mathbf{v}$	$\dfrac{\partial v_j}{\partial x_i}\varepsilon_{ijk}$
Laplacian of a scalar field	$\nabla^2 s$	$\delta_{ij}\dfrac{\partial}{\partial x_i}\left(\dfrac{\partial s}{\partial x_j}\right)$
Laplacian of a vector field	$\nabla^2 \mathbf{v}$	$\delta_{ij}\dfrac{\partial}{\partial x_i}\left(\dfrac{\partial v_k}{\partial x_j}\right)$
Gradient of a vector field	grad $\mathbf{v}$	$\dfrac{\partial v_i}{\partial x_j}$

Table 2.4 Operations with Gradient Operator and Several Components in Cylindrical Coordinates

Operations	Components
grad s	$\dfrac{\partial s}{\partial r}\mathbf{e}_r + \dfrac{1}{r}\dfrac{\partial s}{\partial \theta}\mathbf{e}_\theta + \dfrac{\partial s}{\partial z}\mathbf{e}_z$
div $\mathbf{v}$	$\dfrac{\partial v_r}{\partial r} + \dfrac{v_r}{r} + \dfrac{1}{r}\dfrac{\partial v_\theta}{\partial \theta} + \dfrac{\partial v_z}{\partial z}$
$\nabla^2 s$	$\dfrac{1}{r}\dfrac{\partial}{\partial r}\left(r\dfrac{\partial s}{\partial r}\right) + \dfrac{1}{r^2}\dfrac{\partial^2 s}{\partial \theta^2} + \dfrac{\partial^2 s}{\partial z^2}$
$\nabla^2 \mathbf{v}$	$\left(\dfrac{\partial}{\partial r}\left(\dfrac{1}{r}\dfrac{\partial(rv_r)}{\partial r}\right) + \dfrac{1}{r^2}\dfrac{\partial^2 v_r}{\partial \theta^2} - \dfrac{2}{r^2}\dfrac{\partial v_\theta}{\partial \theta} + \dfrac{\partial^2 v_r}{\partial z^2}\right)\mathbf{e}_r +$ $\left(\dfrac{\partial}{\partial r}\left(\dfrac{1}{r}\dfrac{\partial(rv_\theta)}{\partial r}\right) + \dfrac{1}{r^2}\dfrac{\partial^2 v_\theta}{\partial \theta^2} + \dfrac{2}{r^2}\dfrac{\partial v_r}{\partial \theta} + \dfrac{\partial^2 v_\theta}{\partial z^2}\right)\mathbf{e}_\theta +$ $\left(\dfrac{1}{r}\dfrac{\partial}{\partial r}\left(r\dfrac{\partial v_z}{\partial r}\right) + \dfrac{1}{r^2}\dfrac{\partial^2 v_z}{\partial \theta^2} + \dfrac{\partial^2 v_z}{\partial z^2}\right)\mathbf{e}_z$

Contd...

grad $\mathbf{v}$	$\dfrac{\partial v_r}{\partial r}\mathbf{e}_r\mathbf{e}_r + \left(\dfrac{1}{r}\dfrac{\partial v_r}{\partial \theta} - \dfrac{v_r}{r}\right)\mathbf{e}_r\mathbf{e}_\theta + \dfrac{\partial v_r}{\partial z}\mathbf{e}_r\mathbf{e}_z +$
	$\dfrac{\partial v_\theta}{\partial r}\mathbf{e}_\theta\mathbf{e}_r + \left(\dfrac{1}{r}\dfrac{\partial v_\theta}{\partial \theta} + \dfrac{v_r}{r}\right)\mathbf{e}_\theta\mathbf{e}_\theta + \dfrac{\partial v_\theta}{\partial z}\mathbf{e}_\theta\mathbf{e}_z +$
	$\dfrac{\partial v_z}{\partial r}\mathbf{e}_r\mathbf{e}_z + \dfrac{1}{r}\dfrac{\partial v_z}{\partial \theta}\mathbf{e}_\theta\mathbf{e}_z + \dfrac{\partial v_z}{\partial z}\mathbf{e}_z\mathbf{e}_z$

Table 2.5 Operations with Gradient Operator and Several Components in Spherical Coordinates

Operations	Components
grad s	$\dfrac{\partial s}{\partial r}\mathbf{e}_r + \dfrac{1}{r}\dfrac{\partial s}{\partial \theta}\mathbf{e}_\theta + \dfrac{1}{r\sin\theta}\dfrac{\partial s}{\partial \varphi}\mathbf{e}_\varphi$
div $\mathbf{v}$	$\dfrac{1}{r^2}\dfrac{\partial\left(r^2 v_r\right)}{\partial r} + \dfrac{1}{r\sin\theta}\dfrac{\partial\left(\sin\theta\, v_\theta\right)}{\partial \theta} + \dfrac{1}{r\sin\theta}\dfrac{\partial v_\varphi}{\partial \varphi}$
$\nabla^2 s$	$\dfrac{1}{r^2}\dfrac{\partial}{\partial r}\left(r^2\dfrac{\partial s}{\partial r}\right) + \dfrac{1}{r^2\sin\theta}\dfrac{\partial}{\partial \theta}\left(\sin\theta\dfrac{\partial s}{\partial \theta}\right) + \dfrac{1}{r^2\sin^2\theta}\dfrac{\partial^2 s}{\partial \varphi^2}$
$\nabla^2 \mathbf{v}$	$\left(\dfrac{\partial}{\partial r}\left(\dfrac{1}{r^2}\dfrac{\partial\left(r^2 v_r\right)}{\partial r}\right) + \dfrac{1}{r^2\sin\theta}\dfrac{\partial}{\partial \theta}\left(\sin\theta\dfrac{\partial v_r}{\partial \theta}\right) + \dfrac{1}{r^2\sin^2\theta}\dfrac{\partial^2 v_r}{\partial \varphi^2}\right.$ $\left.-\dfrac{2}{r^2\sin\theta}\dfrac{\partial}{\partial \theta}\left(\sin\theta\, v_\theta\right) - \dfrac{2}{r^2\sin\theta}\dfrac{\partial v_\varphi}{\partial \theta}\right)\mathbf{e}_r +$ $\left(\dfrac{1}{r^2}\dfrac{\partial}{\partial r}\left(r^2\dfrac{\partial v_\theta}{\partial r}\right) + \dfrac{1}{r^2}\dfrac{\partial}{\partial \theta}\left(\dfrac{1}{r^2\sin\theta}\dfrac{\partial}{\partial \theta}\left(\sin\theta\, v_\theta\right)\right)\right.$ $\left.+\dfrac{1}{r^2\sin^2\theta}\dfrac{\partial^2 v_\theta}{\partial \varphi^2} + \dfrac{2}{r^2}\dfrac{\partial v_r}{\partial \theta} - \dfrac{2\cot\theta}{r^2\sin\theta}\dfrac{\partial v_\varphi}{\partial \varphi}\right)\mathbf{e}_\theta +$ $\left(\dfrac{1}{r^2}\dfrac{\partial}{\partial r}\left(r^2\dfrac{\partial v_\varphi}{\partial r}\right) + \dfrac{1}{r^2}\dfrac{\partial}{\partial \theta}\left(\dfrac{1}{r^2\sin\theta}\dfrac{\partial}{\partial \theta}\left(\sin\theta\, v_\varphi\right)\right)\right.$ $\left.+\dfrac{1}{r^2\sin^2\theta}\dfrac{\partial^2 v_\varphi}{\partial \varphi^2} + \dfrac{2}{r^2\sin\theta}\dfrac{\partial v_r}{\partial \varphi} + \dfrac{2\cot\theta}{r^2\sin\theta}\dfrac{\partial v_\theta}{\partial \varphi}\right)\mathbf{e}_\varphi$
grad $\mathbf{v}$	$\dfrac{\partial v_r}{\partial r}\mathbf{e}_r\mathbf{e}_r + \left(\dfrac{1}{r}\dfrac{\partial v_r}{\partial \theta} + \dfrac{v_r}{r}\right)\mathbf{e}_r\mathbf{e}_\theta + \dfrac{\partial v_r}{\partial \varphi}\mathbf{e}_r\mathbf{e}_\varphi +$ $\dfrac{\partial v_\theta}{\partial r}\mathbf{e}_\theta\mathbf{e}_r + \dfrac{1}{r}\dfrac{\partial v_\theta}{\partial \theta}\mathbf{e}_\theta\mathbf{e}_\theta + \dfrac{\partial v_\theta}{\partial \varphi}\mathbf{e}_\theta\mathbf{e}_\varphi +$ $\dfrac{\partial v_\varphi}{\partial r}\mathbf{e}_r\mathbf{e}_\varphi + \dfrac{\partial v_\varphi}{\partial \theta}\mathbf{e}_\theta\mathbf{e}_\varphi + \dfrac{\partial v_\varphi}{\partial \varphi}\mathbf{e}_\varphi\mathbf{e}_\varphi$

The Gauss's divergence theorem is used in formulating the balance laws described in Chapter 3. For a closed surface S containing volume V, Gauss's theorem is given by,

$$\int_S \mathbf{C} \cdot \mathbf{n} dS = \int_V (\operatorname{div} \mathbf{C}) dV . \tag{2.4.1}$$

where $\mathbf{C}$ is a vector or tensor field $\mathbf{n}$ is the unit normal vector to the surface at a given point. This theorem is useful for interchanging between the behaviour of fields on surfaces and inside volumes. Examples of such fields are velocity and stress tensor.

SUMMARY

In this chapter, we reviewed preliminary mathematical concepts and identities that are required for development of governing equations and constitutive relations in continuum mechanics. After defining tensors and their properties, we described several operations in which they are involved. Finally, we also observed derivative operations with the tensors.

EXERCISE

2.1 In Section 2.3.1, we defined invariants of a tensor (I_1, I_2 and I_3) in terms of eigenvalues of a tensor, λ_1, λ_2 and λ_3. Show that the invariants can also be written in terms of the components of a tensor, T_{ij}. (Eqn. (2.3.12):

$$I_T = \sum_i T_{ii}, \quad \text{II}_T = \sum_{i,j} T_{ij} T_{ji}, \quad \text{III}_T = \sum_{i,j,k} T_{ij} T_{jk} T_{ki}$$

Hint: Expand the determinant given in Eqn. (2.3.10). The result will be a cubic equation in λ. The coefficients of λ^2, λ^1 and λ^0 will be the above invariants.

2.2. Using index notation, prove the following identities:

(a) $\mathbf{a} \cdot (\mathbf{b} \times \mathbf{c}) = \mathbf{b} \cdot (\mathbf{c} \times \mathbf{a})$

(b) $(\mathbf{a} \times \mathbf{b}) \times \mathbf{c} = (\mathbf{a} \cdot \mathbf{c}) \mathbf{b} - (\mathbf{b} \cdot \mathbf{c}) \mathbf{a}$

(c) $\operatorname{curl}(\mathbf{a} \times \mathbf{b}) = (\mathbf{b} \cdot \operatorname{grad}) \mathbf{a} - (\mathbf{a} \cdot \operatorname{grad}) \mathbf{b} + \mathbf{a} (\operatorname{div} \mathbf{b}) - \mathbf{b} (\operatorname{div} \mathbf{a})$

(d) $\nabla^2 \mathbf{v} = \operatorname{grad}(\operatorname{div} \mathbf{v}) - \operatorname{curl}(\operatorname{curl} \mathbf{v})$

(e) For a symmetric tensor $\mathbf{T}$ and a vector $\mathbf{v}$, show that $\mathbf{T} \cdot \mathbf{v} = \mathbf{v} \cdot \mathbf{T}$.

2.3. (*a*) Find the double dot product of the following two tensors:

$$\tau = \begin{bmatrix} -p & \dfrac{4\mu\langle v_z\rangle r}{R^2} & 0 \\[2mm] \dfrac{4\mu\langle v_z\rangle r}{R^2} & -p & 0 \\[2mm] 0 & 0 & -p \end{bmatrix}$$

$$\mathbf{D} = \begin{bmatrix} 0 & \dfrac{-4\langle v_z\rangle r}{R^2} & 0 \\[2mm] \dfrac{-4\langle v_z\rangle r}{R^2} & 0 & 0 \\[2mm] 0 & 0 & 0 \end{bmatrix}$$

(*b*) What are the three invariants of **D**?

*Note: The above expressions for tensors τ (deviatoric stress tensor) and **D** (stretching tensor) are encountered in one dimensional fluid flow through pipes. The double dot product defined in (a) above, refers to the rate of work (viscous dissipation) associated with the flow.*

2.4. A tensor **T** transforms every vector into its mirror image with respect to the plane whose normal is $\mathbf{n} = \sqrt{2}/2(\mathbf{e}_1 + \mathbf{e}_2)$.

(*a*) Find the components of **T**

(*b*) Use this linear transformation to find the mirror image of a vector $\mathbf{a} = \mathbf{e}_1 + \mathbf{e}_2$

2.5. Find the tensor **T** which transforms any vector **a** into a vector $\mathbf{b} = \mathbf{m}(\mathbf{a} \cdot \mathbf{n})$ where $\mathbf{m} = \sqrt{2}/2(\mathbf{e}_1 + \mathbf{e}_2)$ and $\mathbf{n} = \sqrt{2}/2(-\mathbf{e}_1 + \mathbf{e}_3)$. Write this tensor as a dyadic product.

2.6. Given $\mathbf{S} = \begin{bmatrix} 1 & 0 & 2 \\ 0 & 1 & 2 \\ 3 & 0 & 3 \end{bmatrix}$, evaluate

(*a*) S_{ij}

(*b*) $S_{ij}S_{jk}$

(*c*) $S_{jk}S_{jk}$

(*d*) $S_{mn}S_{nm}$

2.7. Given the following matrices:

$$\left[\mathbf{a}_{ij}\right] = \begin{bmatrix} 1 \\ 0 \\ 2 \end{bmatrix}, \left[\mathbf{B}_{ij}\right] = \begin{bmatrix} 2 & 3 & 0 \\ 0 & 5 & 1 \\ 0 & 2 & 1 \end{bmatrix}, \left[\mathbf{C}_{ij}\right] = \begin{bmatrix} 0 & 3 & 1 \\ 1 & 0 & 2 \\ 2 & 4 & 3 \end{bmatrix}$$

Demonstrate the equivalence of the following equations and the corresponding matrix equations.

(i) $s = B_{ij}a_i a_j$, $s = \mathbf{a}^T \mathbf{B} \mathbf{a}$

(ii) $b_i = B_{ij}a_j$, $\mathbf{b} = \mathbf{B}\mathbf{a}$

(iii) $C_j = B_{ji}a_i$, $\mathbf{c} = \mathbf{B}\mathbf{a}$

(iv) $D_{ik} = B_{ij}C_{jk}$, $\mathbf{D} = \mathbf{B}\mathbf{C}$

(v) $D_{ik} = B_{ij}C_{kj}$, $\mathbf{D} = \mathbf{B}\mathbf{C}^T$

2.8. Given that $\sigma_{ij} = 2\mu e_{ij} + \lambda\left(e_{kk}\right)\delta_{ij}$, show that

(a) $W = \dfrac{1}{2}\sigma_{ij}e_{ij} = \mu e_{ij}e_{ij} + \dfrac{\lambda}{2}\left(e_{kk}\right)^2$

(b) $P = \sigma_{ij}\sigma_{ij} = 4\mu^2 e_{ij}e_{ij} + \left(e_{kk}\right)^2\left(4\mu\lambda + 3\lambda^2\right)$

Note: The above equation relating $\boldsymbol{\sigma}$ and $\mathbf{e}$ is a stress ($\boldsymbol{\sigma}$)-strain ($\mathbf{e}$) relationship popularly used in linear elasticity. Where μ and λ are called Lame's constants and W is the strain energy density function.

■■■

CHAPTER

3

Continuum Mechanics
Concepts

मया ततमिदं सर्वं ज्गदव्यक्तमूर्तिना । (भगवद्गीता)
(*Mayā tatamidaṃ sarvam jagadavyaktamūatinā* (*Bhagavadgitā*))

By me who is formless in this universe pervaded.

3.1 INTRODUCTION

In Chapter 1, we referred to two types of models *viz*., systems models and material models. A systems model is a mathematical description of the system in which a material exists and the material model is the mathematical description of the material response. The systems model essentially is a conglomeration of the mathematical descriptions of the physical laws that govern the system. This is normally in various mathematical forms such as integral and differential equations. These equations need to be solved within the constraints of boundary and initial conditions, in order to describe the system response in any field situation.

The physical laws that govern a material response are already familiar to us in different contexts. For example, in solid mechanics, we are aware of *equilibrium equations, compatability conditions* and *stress-strain laws* that are common to all solids. Similarly, in fluid mechanics, we are aware of *balances of mass and linear momentum* for Newtonian fluids. Also, we are aware of the *heat conduction equation* from subjects like heat transfer. We also know of the existence of the *laws of thermodynamics*, which are discussed within the contexts such as power generation, refrigeration and phase equilibrium. Very often, we learn each of these subjects in isolation and are unable to see their mutual linkages. For example, it is possible to see that the heat conduction equation and the first law of thermodynamics are manifestations of a statement of balance of energy. Further, these subjects are developed for certain class of materials and applications, where it is sufficient to work with simplified assumptions such as linearity, small deformations, etc.

Since each of the subjects outlined above is relating some aspect of material behavior in different contexts, it is possible to integrate all of these subjects within a common framework. This is attempted in *continuum mechanics*. Within the framework of continuum mechanics, we visualize the description of solid-like and fluid-like behaviour of a material as manifestations of the same set of governing physical laws. We further develop a framework that allows us to see the linkages between concepts of thermodynamics, momentum and energy balances in any material. This framework is developed to be valid within the context of any arbitrary motion. Hence, our first task is to describe this motion. A general description of the motion of any material is called as *kinematics* and this will be developed in the following section.

3.2. KINEMATICS

A space of real numbers is called *real coordinate space* and denoted by $\mathcal{R}$. It is important to note that all the properties that are associated with any material are defined only in $\mathcal{R}$, in which we perform all mathematical operations. $\mathcal{R}$ together with definitions of distance measures and operations with distances (such as dot product, to define angles) is called *Euclidean space $\mathcal{E}$*. It is reasonable to assume that any material consists of a number of *material particles p*, which have various properties like a density, position, velocity, acceleration, etc. in $\mathcal{E}$. However, the material particles themselves *exist* as a set in an *abstract space*, $\boldsymbol{p}$.

Let us consider any body $\mathcal{B}$ as a set of such material particles that exist in an abstract space, having a boundary. If these particles are visualized in a real space, they are said to be in a manifested form, having a particular *configuration*. There is a unique mapping $\kappa(p)$ of any particle p, to its manifestation which we will denote as x_2 in the real space. This transformation is conceptually shown in Fig. 3.1. Thus, we can say that

$$\mathbf{x}' = \kappa'(p). \tag{3.2.1}$$

Let us say that all the representations $\mathbf{x}'$ of the particles occupy a region B' in the real space.

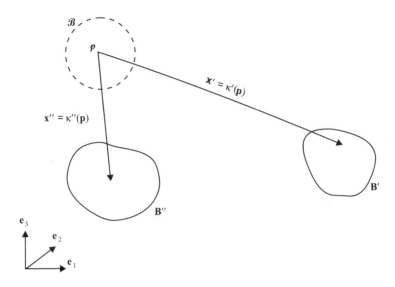

Fig. 3.1 Body $\mathcal{B}$ and its two possible configurations, B′ and B″

We could postulate the existence of another configuration of the body **B** occupying different regions through different mapping functions. For example, Fig. 3.1 also shows a region B″ with a manifestation **x″** through another mapping function $\kappa''(p)$. Hence, we have

$$\mathbf{x}'' = \kappa''(p). \tag{3.2.2}$$

It is convenient to use time τ as a parameter that will monitor the various manifestations of the particle p. To account for this, henceforth we will introduce time τ as a parameter on which the manifestations vary. Various manifestations such of a particle p at different instants of time are denoted as follows:

$$\mathbf{x}'(p, t') \text{ at } \tau = t',$$
$$\mathbf{x}''(p, t'') \text{ at } \tau = t'', \quad \text{etc.} \tag{3.2.3}$$

We can define the *current time* (time at which all observations and calculations are made) and denote this time as $\tau = t$. The manifestation of the particle p at this time is denoted as $\mathbf{x}(p, t)$ and is called as the *current configuration*. Since we now know that there is a one to one correlation between any particle p and its manifestation **x**, it is possible for us to drop p in any further description of the manifestations. Henceforth, we will denote $\mathbf{x}'(p, t')$ as simply $\mathbf{x}'(t')$. Since a manifestation such as $\mathbf{x}'(t')$ is always defined in the real space, we can say that it is the place occupied by p in the real space. This place can be specified as a *coordinate* using an arbitrary origin. Henceforth, we will refer to $\mathbf{x}'(t')$ as a coordinate.

In order to relate the various configurations to each other, it is useful to define a *reference configuration* B^r at time t^r as shown in Fig. 3.2. A reference configuration is a configuration with respect to which all the other configurations are compared. The co-ordinate of the particle p in this reference configuration at time t^r is defined as $\mathbf{X}(t^r)$. It is important to note that at this stage, a reference configuration is a convenient nomenclature given to a chosen configuration and we are not making any statement with regard to its properties such as invariance with time etc. Later, we will find that it may be useful to keep this reference configuration fixed in time, for convenience of certain measurements in some situations. We may also choose this reference configuration to be the current configuration for the ease of handling of motions in many other situations.

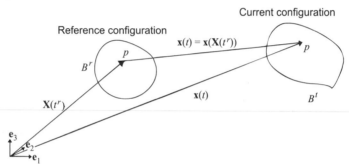

Fig. 3.2 Definitions of reference (undeformed) and current (deformed) configurations and their relation

We recollect at this stage the popular definitions of concepts like motion and deformation. We understand motion of a body or of a material particle, as a movement from one configuration to another. The motion of a body is always defined with reference to any two particles in the body. The motion of a body may be of two types, *viz.*, rigid body motions and deformable motion. Conventionally, a rigid body motion is a motion in which, the distance between any two material particles in the body remains constant as the body undergoes motion. In contrast to this, we have deformable motions where the distance between any two material particles changes. Popularly, the deformable motions are also called as *deformations* taking place in a body.

A particle in the two configurations B^r and B^t can be mapped with respect to each other using a unique mapping function $\hat{x}\,(\mathbf{X}(t^r))$ (see Fig. (3.2)) defined below:

$$\mathbf{x}(t) = \hat{x}\,(\mathbf{X}(t^r)). \qquad (3.2.3)$$

In the following discussions, we will assume that $\hat{x}\,(\mathbf{X})$ is differentiable in $\mathcal{R}$, so that it is possible for us to define a *deformation gradient* tensor $\mathbf{F}$ which is given by

$$\mathbf{F}(\mathbf{X}(t^r), t) = \text{Grad }\hat{x}\ (\mathbf{X}(t^r)) = \nabla\ \hat{x}\ (\mathbf{X}(t^r))$$

$$= \frac{\partial \hat{x}(\mathbf{X}(t^r))}{\partial \mathbf{X}} \tag{3.2.4}$$

We note here that we have used a capital G in the definition of the gradient operation. This is to denote that the gradient is being evaluated with respect to the reference configuration $\mathbf{X}$. The deformation gradient $\mathbf{F}$ is a useful quantity which can define motion or deformation of a body. It is a measure of the change of any current configuration with respect to a reference configuration and is associated with deformation. The evaluation of the deformation gradient is illustrated in Example 3.2.1.

Example 3.2.1: Consider a deformation that is given by the following transformation: $x_1 = X_1 + kX_2$; $x_2 = X_2$ and $x_3 = X_3$. Sketch the deformations associated with this transformation in the x_1– x_2 plane for any arbitrary value of k.

Solution:

We know from the definition of deformation gradient that $F_{ij} = \dfrac{\partial x_i}{\partial X_j}$.

Applying this expression for the transformations defined in the current example, we get $F_{11} = 1$, $F_{12} = k$, $F_{22} = 1$, $F_{33} = 1$ and all other $F_{ij} = 0$. Hence, F for this problem, is defined as:

$$\mathbf{F} = \begin{bmatrix} 1 & k & 0 \\ 0 & 1 & 0 \\ 0 & 0 & 1 \end{bmatrix}$$

Fig. E3.2.1(a) shows a representation of a rectangular element in reference configuration. Fig. E3.2.1(b) represents the same element in the current configuration. As we can clearly see, the current set of transformations represent a shear deformations (that are familiar to us in elementary solid mechanics) in the x_1– x_2 plane.

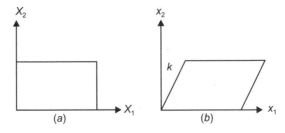

Fig. E3.2.1 Representations of a rectangular element in

(a) reference configuration and (b) current configuration

3.2.1 Transformations

The mapping of points that were outlined in the previous section forms the building block for mapping of a set of points. These sets of points could be in the form a line element, volume element or an area element. Very often, transformation of these sets is very useful in engineering, since we are interested in quantities like stretching distortions that take place in any body, as a consequence of the applied external loads. In this section, we will develop the relations that will map line, volume and area elements between any two configurations of the body.

3.2.1.1 Transformation of Line Elements

As we know, a line element is a set of points that lie in a straight line between two points of interest. In continuum mechanics, line elements are normally used to trace the stretch and orientation of material fibres. Hence, it is useful to look closely at the transformations in a line element. This is done in this section.

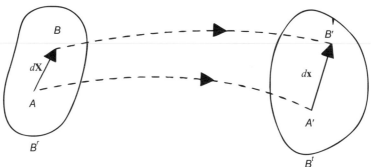

Fig. 3.3 Deformation of a line element

Let us consider two points A and B (representing points p_1 and p_2) in the configuration B^r (as shown in Fig. 3.3), which have co-ordinates $\mathbf{X}$ and $\mathbf{X} + d\mathbf{X}$ respectively, so that the vector joining points A and B in B^r is $d\mathbf{X}$. Since there is a unique mapping between every point $\mathbf{X}$ in the configuration B^r and a corresponding point $\mathbf{x}$ in the current configuration B^t, the points A and B will also have their corresponding representation in B^t as A' (with co-ordinate $\mathbf{x}$), and B' (with coordinates $\mathbf{x} + d\mathbf{x}$). Given these transformations, we can easily write

$$d\mathbf{x} = \hat{x}(\mathbf{X} + d\mathbf{X}) - \hat{x}(\mathbf{X})$$

$$= \hat{x}(\mathbf{X}) + \mathbf{F}(\mathbf{X})d\mathbf{X} + \mathrm{O}(|d\mathbf{X}|^2) - \hat{x}(\mathbf{X}). \quad (3.2.5)$$

Hence,

$$d\mathbf{x} = \mathbf{F}(\mathbf{X})d\mathbf{x} \quad (3.2.6)$$

Equation (3.2.6) indicates that there is a one to one correspondence between fibres $d\mathbf{X}$ in the reference configuration and fibres $d\mathbf{x}$ in the current configuration. This correspondence would require that the deformation gradient $\mathbf{F}$ exists and is non-singular. This means that the Jacobian J (determinant of $\mathbf{F}$) should not vanish, *i.e.*,

$$J = \det \mathbf{F} \neq 0. \tag{3.2.7}$$

The Jacobian defined in Eqn. (3.2.7) has a special significance in kinematic relations. It ensures that all mappings between two configurations are unique. A vanishing Jacobian could imply that the mappings are non-unique. Obviously, we would try to avoid such transformations to take place in a continuous field.

3.2.1.2 Transformation of Volume Elements

Volume elements are necessary to define the region that a material occupies in real space. These elements are used to quantify variables like mass, momentum and energy in a body. Mapping between two volume elements is useful to trace the distortions, expansion, contractions, etc. in any body. The following discussion develops this mapping.

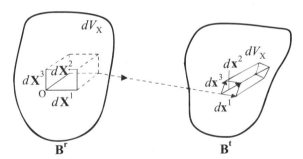

Fig. 3.4 Deformation of a volume element

Consider three line elements $d\mathbf{X}^1$, $d\mathbf{X}^2$ and $d\mathbf{X}^3$ in the reference configuration B^r, all originating at a point O, and forming a right hand system, as shown in Fig. 3.4. It may be noted that these three line elements will define a volume dV_X that is given by the expression

$$dV_X = (d\mathbf{X}^1 \times d\mathbf{X}^2) \cdot d\mathbf{X}^3 \tag{3.2.8}$$

Each of the line elements $d\mathbf{X}^1$, $d\mathbf{X}^2$ and $d\mathbf{X}^3$, can be transformed into corresponding line elements in the current configuration B^t, in which they can be denoted by $d\mathbf{x}^1$, $d\mathbf{x}^2$ and $d\mathbf{x}^3$ using the transformations defined in Eqn. (3.2.6). Further, these line elements themselves will define a volume element dV_x, that is defined by

$$dV_x = (d\mathbf{x}^1 \times d\mathbf{x}^2) \cdot d\mathbf{x}^3 \tag{3.2.9}$$

A relationship between the volume elements defined in Eqns. (3.2.8) and (3.2.9) can be obtained by using Eqn. (3.2.6), as shown below:

$$dV_x = \left(d\mathbf{x}^1 \times d\mathbf{x}^2 \right) \cdot d\mathbf{x}^3$$

$$= \left[\left(\mathbf{F} d\mathbf{X}^1 \right) \times \left(\mathbf{F} d\mathbf{X}^2 \right) \right] \cdot \left(\mathbf{F} d\mathbf{X}^3 \right)$$

$$= \left(\det \mathbf{F} \right) \left[\left(d\mathbf{X}^1 \times d\mathbf{X}^2 \right) \cdot d\mathbf{X}^3 \right] \quad (3.2.10)$$

Hence,

$$dV_x = J dV_X \quad (3.2.11)$$

The transformation relation between volume elements defined in Eqn. (3.2.11) will be useful when we want to relate *field quantities* (quantities that are defined at any point) that are defined in any configuration to the same quantities defined in the new configuration. They will be used in defining any integral relation in continuum mechanics as will be seen in the later part of this chapter.

3.2.1.3 Transformation of Area Elements

Area elements are necessary to define all physical quantities that act on material surfaces. Typically these quantities are in the form of quantities like heat fluxes, stresses etc. It would be of interest to track changes in any area element defined in a body, even as the body moves from one configuration to another. This tracking is done through an area mapping, which will be developed in this section.

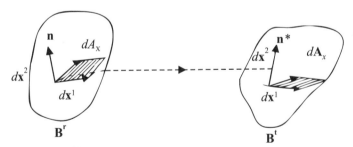

Fig. 3.5 Deformation of an area element

Areas are always associated with a vector $\mathbf{n}$ which defines the orientation of the normal to the surface. We can again use the line elements to define areas at any given point. Areas will get defined as the cross products of two line elements that are acting at any given point. Defining dA_X and dA_x respectively as the magnitudes of the areas in the two configurations, we can define

$$\text{and } d\mathbf{X}^1 \times d\mathbf{X}^2 = dA_X \, \mathbf{n} \text{ and } d\mathbf{x}^1 \times d\mathbf{x}^2 = dA_x \, \mathbf{n}^* \quad (3.2.12)$$

where $\mathbf{n}$ represents the unit normal to the area element in the reference configuration and $\mathbf{n}^*$ refers to the unit normal to the element in the current configuration as shown in Fig. 3.5. Using the definitions of volume elements in Eqns. (3.2.8) and (3.2.9), we get

$$dV_x = \left(d\mathbf{x}^1 \times d\mathbf{x}^2\right) \cdot d\mathbf{x}^3 = \left((dA_x \ \mathbf{n}^*) \cdot \mathbf{F}\right) d\mathbf{X}^3 \text{ and} \qquad (3.2.13)$$

$$dV_X = (dA_X \ \mathbf{n}) \cdot d\mathbf{X}^3 \qquad (3.2.14)$$

Using Eqns. (3.2.12), (3.2.13) and (3.2.14), we get

$$\left((dA_x \ \mathbf{n}^*) \cdot \mathbf{F}\right) d\mathbf{X}^3 = J \ (dA_X \ \mathbf{n}) \cdot d\mathbf{X}^3 . \qquad (3.2.15(a))$$

Using the rule of scalar triple product of tensors, we get

$$dA_x \left(\mathbf{F}^T \ \mathbf{n}^*\right) \cdot d\mathbf{X}^3 = J \ (dA_X \ \mathbf{n}) \cdot d\mathbf{X}^3$$

$$\Rightarrow \left(dA_x \left(\mathbf{F}^T \ \mathbf{n}^*\right) - J \ dA_X \ \mathbf{n}\right) \cdot d\mathbf{X}^3 = 0 . \qquad (3.2.15(b))$$

Hence,

$$dA_x \ \mathbf{F}^T \mathbf{n}^* = J \ dA_X \ \mathbf{n}$$

or $\qquad\qquad dA_x \ \mathbf{n}^* = J \ dA_X \ (\mathbf{F}^T)^{-1} \ \mathbf{n} \qquad (3.2.15(c))$

Taking the determinant on both sides of Eqn. (3.2.15), we get

$$dA_x = J \ dA_X \ |(\mathbf{F}^T)^{-1} \ \mathbf{n} \ |. \qquad (3.2.16)$$

Equation (3.2.16) is a useful relation that maps the areas in two configurations. We find that this equation also relates the area-normals in the two configurations and shows that the deformation gradient is necessary to establish this mapping. This equation will be used later when we define force measures called *tractions*.

3.2.2 Important Types of Motions

The transformation of line, volume and area elements that have been developed in the previous sections will now be used to examine transformations in three of the most commonly encountered motions. Even though the transformation from the reference configuration to the current configurations could be in any arbitrary form, it would help us to examine some special types of deformations. Any other arbitrary deformations could then be visualized in terms of these special deformations. It should be noted that only a sub-region D^r of B^r is being considered to undergo a specific type of deformation. The reference configuration, in the following discussion, is referred to as the *undeformed configuration*, while the current configuration is referred to as *deformed configuration*.

3.2.2.1 Isochoric Deformations

Consider a region D^r in the reference configuration B^r of the body $\mathcal{B}$. All particles within the region D^r will occupy a region D^t in the deformed configuration of the body as shown in Fig. 3.6. Let us assume that deformation is volume preserving, *i.e.*, the total volume in the region D^r is the same as the total region in the region D^t. This really means that material, even though undergoing some changes in shapes, will not expand or contract. Hence, we can say that (using Eqn. (3.2.11)),

$$\int_{D^r} dV_X = \int_{D^t} dV_x = \int_{D^r} J\, dV_X\,.$$

(3.2.17)

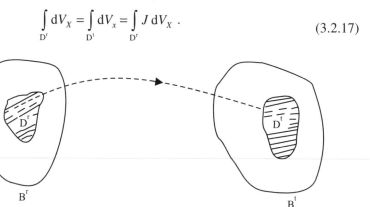

Fig. 3.6 Definition of sub-region within a body

On rearranging, we get

$$\int_{D^v} (J-1)\, dV_X = 0 \qquad \text{for all parts of } D^r$$

(3.2.18)

Therefore, it follows that for isochoric deformations,

$$J = 1.$$

(3.2.19)

In practice, we find that there are two types of deformations that are commonly used to characterize material deformation. One of them is *extension* and the other is *shear*. Extension is a deformation in which the diagonal components of the deformation gradient are non-unity and the off-diagonal components are zeros. The three diagonal components signify stretches in three mutually perpendicular directions. Shear is a deformation in which the diagonal components of the deformation gradient are unity and at least one of the non-diagonal components is non-zero. If only one of the non-diagonal components of the deformation gradient is zero, the deformation is called as a simple shear. The following example shows some more details of a simple shear deformation.

Example of Isochoric Deformation in Simple Shear

Consider the deformations of simple shear that are given in Example 3.2.1 (Fig. E3.2.1). The transformations given in this example are recollected here for reference:

$$x_1 = X_1 + k X_2, \ x_2 = X_2 \text{ and } x_3 = X_3 \qquad (3.2.20)$$

We recollect the deformation gradient tensor F, associated with this deformation as

$$\mathbf{F} = \begin{pmatrix} 1 & k & 0 \\ 0 & 1 & 0 \\ 0 & 0 & 1 \end{pmatrix} = \mathbf{I} + k\mathbf{e}_1\mathbf{e}_2 \qquad (3.2.21)$$

It can be easily seen from Eqn. (3.2.21) that $J = \det \mathbf{F} = 1$. Hence, from Eqn. (3.2.19), we identify that the shear deformations defined in Eqn. (3.2.20) represent isochoric deformations.

3.2.2.2 Rigid Body Motion

A rigid body is normally defined as a body that experiences no deformations. In other words, distance between any two particles say $\mathbf{X}$ and $\mathbf{Z}$ will remain the same, even though the body itself may undergo some motions. This can be mathematically expressed as

$$\left|\hat{x}(\mathbf{Z}) - \hat{x}(\mathbf{X})\right| = |\mathbf{Z} - \mathbf{X}| \ \text{ for all points in the body.}$$
$$(3.2.22)$$

Equation (3.2.22) can also be written as

$$\left(\hat{\mathbf{x}}(\mathbf{Z}) - \hat{\mathbf{x}}(\mathbf{X})\right)\cdot\left(\hat{\mathbf{x}}(\mathbf{Z}) - \hat{\mathbf{x}}(\mathbf{X})\right) = (\mathbf{Z} - \mathbf{X})\cdot(\mathbf{Z} - \mathbf{X}). \qquad (3.2.23)$$

Differentiating Eqn. (3.2.23) w.r.t. $\mathbf{X}$, we get

$$-2\mathbf{F}^T(\mathbf{X})\cdot\left(\hat{\mathbf{x}}(\mathbf{Z}) - \hat{\mathbf{x}}(\mathbf{X})\right) = -2(\mathbf{Z} - \mathbf{X}) \qquad (3.2.24)$$

Differentiating Eqn. (3.2.24) again w.r.t. $\mathbf{Z}$, we get

$$\mathbf{F}^T(\mathbf{X})\,\mathbf{F}(\mathbf{Z}) = \mathbf{I} \qquad (3.2.25)$$

Substituting $\mathbf{Z} = \mathbf{X}$ in Eqn. (3.2.25), we get

$$\mathbf{F}^T(\mathbf{X})\,\mathbf{F}(\mathbf{X}) = \mathbf{F}^T(\mathbf{Z})\,\mathbf{F}(\mathbf{Z}) = \mathbf{I} \qquad (3.2.26)$$

Equation (3.2.26) suggests that $\mathbf{F}(\mathbf{X})$ has to be a proper orthogonal tensor $\mathbf{Q}$, which is normally a tensor associated with pure rotations. Further, we can prove that $\mathbf{Q}$ is independent of the position $\mathbf{X}$. If $\mathbf{Q}$ is independent of $\mathbf{X}$, from the definition of deformation gradient $\mathbf{F}$, we can easily state that rigid body transformations are associated with transformations of the type.

$$\mathbf{x} = \mathbf{Q}\,\mathbf{X} + \mathbf{b}. \qquad (3.2.27)$$

Clearly, rigid transformations of the type $x_1 = X_1 + a$, $x_2 = X_2 + b$, $x_3 = X_3 + c$, where a, b and c are constants, are examples of a rigid body motion.

3.2.2.3 Homogeneous Deformations

Homogeneous deformations are those deformations in which the deformation gradient is not a function of the position $\mathbf{X}$ (and it need not be orthogonal). Hence, the deformations are characterized by transformations of the type

$$\mathbf{x} = \mathbf{F}\,\mathbf{X} + \mathbf{b}. \tag{3.2.28}$$

where $\mathbf{F}$ is a constant (not a function of $\mathbf{X}$). This would mean that straight lines, planes and arbitrarily shaped bodies in the undeformed configuration B^r would retain their respective shapes in the deformed configurations B^t. We can easily see that the shear transformations defined in Eqn. (3.2.20) will result in a constant $\mathbf{F}$ and hence are examples of a homogeneous deformation.

3.2.3 Decomposition of Deformation Gradient

Deformation gradient is rarely used directly, because it is non-zero for rigid body motion. However, we are more interested in deformation measures (with no rigid body motion, *i.e.*, no translation and no rotation). These are also called the stretch measures. In the following section, we will see how deformation gradient is decomposed into two tensors, signifying rotation and stretch.

3.2.3.1 Polar Decomposition Theorem

Polar decomposition theorem states that a non-singular tensor can be decomposed into a positive definite tensor and an orthogonal tensor. Therefore, it is possible to decompose $\mathbf{F}$ into symmetric positive definite tensors $\mathbf{U}$ and $\mathbf{V}$ as well as an orthogonal tensor $\mathbf{R}$ such that

$$\mathbf{F} = \mathbf{RU} = \mathbf{VR} \tag{3.2.29}$$

The tensor $\mathbf{U}$ is called as the right stretch tensor. It represents a stretch that one could give to the body in its underformed configuration *before* it is subjected to a rigid body rotation $\mathbf{R}$ so as to bring it into the current configuration. The tensor $\mathbf{V}$ is called as the left stretch tensor. It represents a stretch that one could give to the body *after* it is subjected to a rigid body rotation $\mathbf{R}$.

Example 3.2.2: Do a polar decomposition of the deformation gradient matrix obtained for simple shear in Example 3.2.1 and obtain the matrices $\mathbf{U}$ and $\mathbf{V}$. Assume $k = 3/2$.

Solution:

Recall that the deformation gradient associated with simple shear in Eqn. (3.2.21) is given by

$$\mathbf{F} = \begin{pmatrix} 1 & k & 0 \\ 0 & 1 & 0 \\ 0 & 0 & 1 \end{pmatrix}.$$

We further note that $\mathbf{F}^T\mathbf{F} = \mathbf{U}^T\mathbf{R}^T\mathbf{R}\mathbf{U} = \mathbf{U}^2$ since $\mathbf{R}$ is an orthogonal tensor. Substituting for $k = 3/2$ in this expression, we get

$$\mathbf{U}^2 = \begin{pmatrix} 1 & 3/2 & 0 \\ 3/2 & 13/4 & 0 \\ 0 & 0 & 1 \end{pmatrix} \qquad \text{(E3.2.1)}$$

We can subject Eqn. (E3.2.1) above through any standard eigenvalue extractor, to obtain the eigenvalues and eigenvectors as

$$\lambda_{U^2} = \begin{pmatrix} 4 \\ 1/4 \\ 1 \end{pmatrix} \text{ and } \mathbf{X}_{U^2} = \begin{pmatrix} 1/\sqrt{5} & 2/\sqrt{5} & 0 \\ -2/\sqrt{5} & 1/\sqrt{5} & 0 \\ 0 & 0 & 1 \end{pmatrix}. \qquad \text{(E3.2.2)}$$

Using the expressions given in Eqn. (E3.2.2) above, it is easy to extract the eigenvalues of $\mathbf{U}$ (as square roots of eigenvalues of $\mathbf{U}^2$) and eigenvectors of $\mathbf{U}$ (using the eigenvectors of $\mathbf{U}^2$) as

$$\mathbf{U} = \begin{pmatrix} 4/5 & 3/5 & 0 \\ 3/5 & 17/10 & 0 \\ 0 & 0 & 1 \end{pmatrix} \qquad \text{(E3.2.3)}$$

Using the expression for $\mathbf{U}$ given in Eqn. (E3.2.3), we can obtain the $\mathbf{R}$ matrix as

$$\mathbf{R} = \mathbf{F}\mathbf{U}^{-1} = \begin{pmatrix} 4/5 & 3/5 & 0 \\ -3/5 & 4/5 & 0 \\ 0 & 0 & 1 \end{pmatrix}. \qquad \text{(E3.2.4)}$$

The above example illustrates the procedure by which the rotation tensor $\mathbf{R}$ and the right stretch tensor $\mathbf{U}$ can be extracted for any deformation gradient $\mathbf{F}$. A similar procedure can be followed to extract the left stretch tensor $\mathbf{V}$ from $\mathbf{F}$ and it can be shown that the $\mathbf{R}$ matrix will remain the same for both the decompositions.

Modelling of Engineering Materials

In Example 3.2.2, it was seen that we had to define a tensor $\mathbf{U}^2$, while extracting $\mathbf{U}$. $\mathbf{U}^2$ is useful in defining many kinematic quantities and is called as the *right Cauchy-Green deformation tensor*. In a similar fashion, it would be useful to define where $\mathbf{B}=\mathbf{V}^2$ is called as the *left Cauchy-Green or Finger deformation tensor*. Therefore,

$$\mathbf{C} = \mathbf{U}^2 \text{ and } \mathbf{B} = \mathbf{V}^2. \qquad (3.2.30)$$

3.2.3.2 Stretches

In most deformations, it is useful to define a measure called *stretch*. *Stretch* is normally understood to be the relative deformation or extension of a line element. Consider a line element of lengths $d\mathbf{X}$ and ds, with an orientation $\mathbf{m}$ as shown in Figure 3.7. Assume that the element $d\mathbf{X}$ gets transformed into $d\mathbf{x}$, whose length is ds^*, with an orientation $\mathbf{m}^*$ as shown in the figure. From the figure, we have

$$d\mathbf{x} = ds^* \, \mathbf{m}^* \text{ and } d\mathbf{X} = ds \, \mathbf{m} \qquad (3.2.31)$$

The stretch or extension ratio λ of the element is defined as

$$\lambda = \frac{ds^*}{ds} \qquad (3.2.32)$$

Using the definitions of $\mathbf{F}$ from Eqn. (3.2.4) and λ from Eqns. (3.2.32) and the expressions for the line elements from Eqn. (3.2.31), we can derive the following expressions for the extension ratio in terms of the right stretch tensor $\mathbf{U}$ and the left stretch tensor $\mathbf{V}$ as

$$\lambda = \sqrt{\left(\mathbf{U}^2 \cdot \mathbf{m}\right) \cdot \mathbf{m}} \; = \; \frac{1}{\sqrt{\left(\mathbf{V}^{-2} \cdot \mathbf{m}^*\right) \cdot \mathbf{m}^*}} \qquad (3.2.33)$$

The stretch ratios defined in this section will form the basis for the definition of strain measures, which are popularly defined in mechanics. These quantities are defined in the next section.

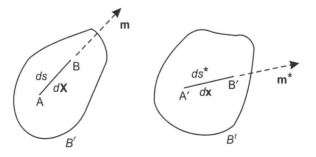

Fig. 3.7 Rotation and stretching of a line element

3.2.4 Strain Measures

Strains may be defined as measures of extension in a material. They are most naturally defined in terms of stretch tensor **U** or **V**. We have observed in above Eqns. (3.2.29) and (3.2.33) that these tensors contain information about stretching in the material. Several strain measures are defined in the literature. **U**, as we have seen in Section 3.2.3.1, is a stretch operating on undeformed configuration of the material. On the other hand, **V** is a stretch that is imposed on the rotated configuration of the material. In the literature, strain measures based on **U** are referred to as the *Lagrangian* measures, while the strain measures based on **V** are referred to as *Eulerian* measures. Some of them are listed in Table 3.1.

Table 3.1 Strain Measures used in the Literature

Strain type	Lagrangian Measure	Eulerian Measure
Green strain (**E**)	$\frac{1}{2}\left(\mathbf{U}^2 - \mathbf{I}\right)$	$\frac{1}{2}\left(\mathbf{I} - \mathbf{V}^{-2}\right)$
Generalized Green strain	$\frac{1}{m}\left(\mathbf{U}^m - \mathbf{I}\right)$	$\frac{1}{m}\left(\mathbf{I} - \mathbf{V}^{-m}\right)$
Hencky strain	$\ln \mathbf{U}$	$\ln \mathbf{V}$

3.2.4.1 Displacements

The motion of any particle in the body can also be described in terms of a quantity called as displacement **u(X)**, that is associated with every particle in the body, referenced with respect to undeformed configuration (see Fig. 3.8). The displacement of any particle is defined as

$$\mathbf{u(X)} = \mathbf{x} - \mathbf{X} = \hat{x}\,(\mathbf{X}) - \mathbf{X}. \qquad (3.2.34)$$

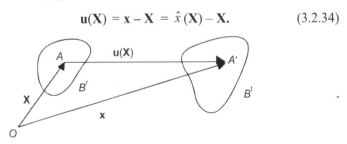

Fig. 3.8 Definition of displacement

It is often also useful to measure the gradient of the displacement with respect to the undeformed configuration. This is defined by the symbol **H(X)**, which is given by the following expression

$$\mathbf{H(X)} = \text{Grad } \mathbf{u} = \mathbf{F} - \mathbf{I} \qquad (3.2.35)$$

The Green Lagrange strain $\mathbf{E}$ can be expressed in terms of the displacement gradient $\mathbf{H}$ as shown below:

$$\mathbf{E(X)} = \frac{1}{2}\left(\mathbf{U}^2 - \mathbf{I}\right) = \frac{1}{2}\left(\mathbf{F}^T\mathbf{F} - \mathbf{I}\right) = \frac{1}{2}\left(\mathbf{H} + \mathbf{H}^T + \mathbf{H}^T\mathbf{H}\right). \qquad (3.2.36)$$

$\mathbf{E}$ can also be written in its component form as

$$E_{ij} = \frac{1}{2}\left(\frac{\partial u_i}{\partial X_j} + \frac{\partial u_j}{\partial X_i} + \frac{\partial u_k}{\partial X_i}\frac{\partial u_k}{\partial X_j}\right) \qquad (3.2.37)$$

3.2.4.2 Infinitesimal Strains

The strain measure ($\mathbf{e}$) and rotation (ω) normally used to describe deformation in theory of elasticity are valid for infinitesimal strain and infinitesimal rotation. These infinitesimal measures can be derived from the right stretch tensor $\mathbf{U}$ and the rotation tensor $\mathbf{R}$, after neglecting the higher order terms. Higher order terms are negligible for infinitismal measures. Thus, we have

$$\mathbf{U} - \mathbf{I} = \frac{1}{2}\left(\mathbf{H} + \mathbf{H}^T\right) + O\left(|\mathbf{H}|^2\right) \qquad (3.2.38)$$

Therefore, for infinitismal deformations, we can define

$$\mathbf{e} = \mathbf{U} - \mathbf{I} = \frac{1}{2}\left(\mathbf{H} + \mathbf{H}^T\right) \qquad (3.2.39)$$

On similar lines, the rotation tensor $\mathbf{R}$ defined in Eqn. (3.2.29) can be used to define the infinitesimal rotation tensor ω as shown below

$$\omega = \mathbf{R} - \mathbf{I} = \frac{1}{2}(\mathbf{H} - \mathbf{H}^T) \qquad (3.2.40)$$

So far we have described spatial derivatives of position vectors, which are helpful in defining quantities such as strains. The spatial derivative operators are with respect to $\mathbf{X}$, the reference coordinate. It is possible to define all the above spatial derivatives with respect to current coordinate $\mathbf{x}$ also, as given in Section 3.2.6. However, these derivatives are more meaningful after a description of the time derivatives, which is given below.

3.2.5 Motions

Motions are descriptions of movements of various particles with time. Hence, motions are invariably associated with time derivatives of any field quantity such as position, density, deformation gradient, etc. Spatial derivatives such as defined in Sections 3.2.3 and 3.2.4, will mostly find their place in any description of motions.

Since most time derivatives are invariably associated with changes in the position co-ordinates of the material points, it is useful to define the following time derivates of the current position co-ordinate of a field point:

$$v(\mathbf{X},t) = \frac{\partial \mathbf{x}}{\partial t} = \frac{\partial \hat{x}(\mathbf{X},t)}{\partial t}$$

$$\mathbf{a}(\mathbf{X},t) = \frac{\partial \mathbf{v}(\mathbf{X},t)}{\partial t}$$

(3.2.41)

where $\mathbf{v}$ and $\mathbf{a}$ are the velocity and acceleration of a material particle, respectively. While a time derivative of any scalar quantity like temperature could take place even without motions, time derivatives are invariably associated with changes in the current position of the field variables of interest. The derivative of any field quantity after including motions is called as total time derivative. Hence, the total derivative of any scalar field quantity say θ is given by

$$\dot{\theta} = \frac{\partial \theta}{\partial t} + \frac{\partial \theta}{\partial x_i} \frac{\partial x_i}{\partial t}$$

(3.2.42)

Denoting $\dfrac{\partial \theta}{\partial t}$ as θ' and $\dfrac{\partial \theta}{\partial x_i}$ as grad θ (gradient of θ), Eqn. (3.2.42) can be rewritten as

$$\dot{\theta} = \theta' + \mathbf{v} \cdot grad\,\theta$$

(3.2.43)

The total time derivative is also denoted by the symbol $\dfrac{D}{Dt}$ in the literature.

Based on the Eqns. (3.2.41) and (3.2.43), it is possible to show that

$$\dot{\mathbf{x}} = \mathbf{v}$$
$$\ddot{\mathbf{x}} = \dot{\mathbf{v}} = \mathbf{a}$$

(3.2.44)

3.2.5.1 Velocity Gradient

The spatial gradient of velocity with respect to the current co-ordinate $\mathbf{x}$ is defined as the velocity gradient, *viz.*

$$\mathbf{L} = grad\ \mathbf{v} = \frac{\partial \mathbf{v}}{\partial \mathbf{x}}$$

(3.2.45)

The velocity gradient is a fundamental kinematic quantity that defines flow in any material. Very often, it is more convenient to define flow in terms of the symmetric part of the velocity gradient $\mathbf{D}$ (also called as the *stretching tensor*) and the anti-symmetric part of the velocity gradient $\mathbf{W}$ (also known as the *spin tensor*). These quantities are related to the velocity gradient **L** as defined below:

$$\mathbf{D} = \frac{1}{2}\left(\mathbf{L} + \mathbf{L}^{\mathrm{T}}\right) \text{ and } \mathbf{W} = \frac{1}{2}\left(\mathbf{L} - \mathbf{L}^{\mathrm{T}}\right). \qquad (3.2.46)$$

It can be easily seen that the following properties are always true for the tensors **D** and **W** :

$$\mathbf{D} = \mathbf{D}^{\mathrm{T}} \text{ and } \mathbf{W} = -\mathbf{W}^{\mathrm{T}}. \qquad (3.2.47)$$

In the literature, **D** is also called the *strain rate tensor* or *rate of strain tensor*, while **W** is also referred to as the *vorticity tensor*.

Example 3.2.3: Show the relationship between the time derivative of the displacement gradient and the velocity gradient **L**.

Solution:

We know that the deformation gradient **F** is defined by, from Eqn. (3.2.4),

$$\mathbf{F} = \frac{\partial \hat{x}(\mathbf{X}, t)}{\partial \mathbf{X}} = \operatorname{Grad} \hat{x}.$$

Hence, we have

$$\dot{\mathbf{F}} = \frac{\partial}{\partial t}(\operatorname{Grad} \hat{x}) = \operatorname{Grad} \frac{\partial \hat{x}}{\partial t} = \operatorname{Grad} \mathbf{v}$$

$$= (\operatorname{grad} \mathbf{v})\mathbf{F} = \mathbf{L}\mathbf{F}.$$

Therefore,

$$\mathbf{L} = \dot{\mathbf{F}}\mathbf{F}^{-1}$$

The above equation tells us that the velocity gradient is proportional to the time derivative of the deformation gradient associated with the material.

Components of the Velocity Gradient

The components of the velocity gradient are better understood if we examine the deformations of single fibres in the material. Consider a fibre d**x** of length ds and oriented along the direction **m**. Since $d\mathbf{x} = \mathbf{F}\,d\mathbf{X}$ and $d\mathbf{X}$ is time independent, we can derive the following expressions for the time rate of change of the elemental fibre d**x**:

$$d\dot{\mathbf{x}} = \dot{\mathbf{F}}\,d\mathbf{X} = \mathbf{L}\mathbf{F}\,d\mathbf{X} = \mathbf{L}d\mathbf{x} \qquad (3.2.48)$$

Since $d\mathbf{x} = ds\,\mathbf{m}$, we get

$$ds^2 = d\mathbf{x} \cdot d\mathbf{x}.$$ (3.2.49(a))

Differentiating the above equation with respect to time, we get

$$ds\,(d\dot{s}) = d\mathbf{x} \cdot d\dot{\mathbf{x}} = (ds)^2\,\mathbf{m} \cdot \mathbf{Lm}$$ (3.2.49(b))

Simplifying, we obtain

$$\frac{d\dot{s}}{ds} = \mathbf{Lm} \cdot \mathbf{m} = \mathbf{Dm} \cdot \mathbf{m}. \quad (\because \mathbf{L} = \mathbf{D} + \mathbf{W} \text{ and } \mathbf{Wm} \cdot \mathbf{m} = 0)$$ (3.2.49(c))

Equation (3.2.49) gives us a simple expression for the unit rate of extension (also sometimes called as the strain rate) of a fibre $d\mathbf{x}$ in the deformed configuration. By picking various directions for $\mathbf{m}$, we could get expressions and physical interpretation for the different components of $\mathbf{D}$. For example, by picking $\mathbf{m}$ to be oriented along the principal direction $\mathbf{e}_1$, we know that D_{11} will represent the relative rate of extension of a fiber along the $\mathbf{e}_1$ direction. It can further be proved (Exercise Problem 3.7) that D_{12} represents the rate of change of the angle between any two fibers that were originally oriented along the principal directions $\mathbf{e}_1$ and $\mathbf{e}_2$.

Rate of Change of Volume

We know that the volume elements in the undeformed configuration (dV_X) and the deformed configurations (dV_x) are related to each other through the Jacobian J in the relation, Eqn. (3.2.11),

$$dV_x = J\,dV_X$$ (3.2.50(a))

Noting that dV_X does not change with time and expanding dV_x in terms of its component line elements, it is possible to prove that

$$\dot{J} = J\,\mathrm{tr}(\mathbf{L}) = J\,\mathrm{tr}(\mathbf{D}) = J\,\mathrm{div}\,\mathbf{v}.$$ (3.2.50(b))

Equation (3.2.50) gives us a good criterion to define special kinds of deformation like volume preserving (isochoric $\dot{j} = 0$) deformations which are defined by

Rigid Body Motions

We have seen earlier that the rigid body transformations are given by transformations of the type given in Eqn. (3.2.27). From this equation, it is possible to define an inverse transformation of the type

$$\mathbf{X} = \mathbf{Q}^T\mathbf{x} - \mathbf{Q}^T\mathbf{b}.$$ (3.2.51)

Based on Eqns. (3.2.27) and (3.2.51), it is possible to arrive at an expression for the velocity $\mathbf{v}\,\dot{\mathbf{x}}$, (Eqn. (3.2.44)), in terms of a generalized body spin tensor $\Omega(t)$ and a space independent tensor $\mathbf{C}(t)$ as (see Exercise 3.10)

$$\mathbf{v} = \Omega(t)\mathbf{x} + \mathbf{C}(t)$$ (3.2.52)

Since $\Omega(t)$ is skew-symmetric, it is always possible to define a rotation vector ω_r which will express Eqn. (3.2.52) through a cross product as defined below:

$$\mathbf{v} = \omega_r(t) \times \mathbf{x} + \mathbf{C}(t) \qquad (3.2.53)$$

Description of rigid body motions of any body through Eqn. (3.2.53) is very useful in describing the total motion in any body and this will be seen more clearly in subsequent sections. It can be easily shown that the following identities are valid for a rigid body motion

$$\mathbf{L} = \Omega \; ; \mathbf{D} = \mathbf{0} \; \text{ and } \mathbf{W} = \Omega. \qquad (3.2.54)$$

It should be noted that $\mathbf{W}$, $\mathbf{L}$ and $\mathbf{D}$ can be expressed in terms of the derivatives of stretch and rotation tensors $\dot{\mathbf{U}}$, $\dot{\mathbf{V}}$ and $\dot{\mathbf{R}}$ (Exercise 3.4).

3.2.6 Relative Deformation Gradient

Deformation gradient defined thus far is based on reference configuration, which is independent of time. However, it is possible to define the deformation gradient as relation between any configuration at arbitrary time (τ) and the present time (t):

$$\mathbf{F}_t(\tau) = \frac{\partial \mathbf{x}(\tau)}{\partial \mathbf{x}(t)} = \frac{\partial \mathbf{x}^\tau}{\partial \mathbf{x}}. \qquad (3.2.55)$$

where $\mathbf{x}(\tau)$ is the configuration at time τ. This deformation gradient is referred to as *relative deformation gradient* and we can show that it is related to the deformation gradient defined earlier (in Eqn. (3.2.4)) as follows:

$$\mathbf{F}_t(\tau) \cdot \mathbf{F}(t) = \mathbf{F}(\tau). \qquad \left(\because \mathbf{F}(t) = \frac{\partial \mathbf{x}}{\partial \mathbf{X}}, \; \mathbf{F}(\tau) = \frac{\partial \mathbf{x}^\tau}{\partial \mathbf{X}} \right) \qquad (3.2.56)$$

This definition of deformation gradient, *i.e.*, the relative deformation gradient is used often by researchers working on fluid materials. As earlier, we can define the relative left and right stretch tensors and the relative rotation tensor with respect to the relative deformation gradient (analogous to Eqn. (3.2.29)):

$$\mathbf{F}_t = \mathbf{R}_t \cdot \mathbf{U}_t = \mathbf{V}_t \cdot \mathbf{R}_t. \qquad (3.2.57)$$

The relative strain measures can then be defined as (analogous to Eqn. (3.2.30))

$$\mathbf{C}_t = \mathbf{U}_t^2$$
$$\mathbf{B}_t = \mathbf{V}_t^2 \qquad (3.2.58)$$

It can be shown that the velocity gradient, stretching and spin tensors are related to the derivatives of relative measures of deformation, as follows

$$\mathbf{L} = \dot{\mathbf{F}}_t(\tau)$$
$$\mathbf{D} = \dot{\mathbf{U}}_t(\tau) = \dot{\mathbf{V}}_t(\tau), \qquad (3.2.59)$$
$$\mathbf{W} = \dot{\mathbf{R}}_t(\tau)$$

Recall that we had shown earlier that $\mathbf{L} = \dot{\mathbf{F}} \cdot \mathbf{F}^{-1}$ (Example 3.2.1). It can be seen from Eqn. (3.2.59) that $\mathbf{L}$ is the rate of change of the relative deformation gradient. Earlier, we had seen that $\mathbf{D}$ and $\mathbf{W}$ are symmetric and skew-symmetric parts of $\mathbf{L}$ (Eqn. (3.2.46)). Based on Eqn. (3.2.59), we can infer that the stretching tensor $\mathbf{D}$ is the rate of change of relative stretch tensor $\mathbf{U}_t$ and $\mathbf{V}_t$. Similarly, the spin tensors is the rate of change of relative rotation tensor.

The derivatives of measures based on deformation gradient can always be related to derivatives of measures based on relative deformation gradient, since the two deformation gradients are related to each other. For example, it can be shown that

$$\dot{\mathbf{U}}_t = \frac{1}{2}\mathbf{R}\left[\dot{\mathbf{U}}\mathbf{U}^{-1} + \mathbf{U}^{-1}\mathbf{U}\right]\mathbf{R}^{\mathrm{T}} = \mathbf{D} \qquad (3.2.60)$$

Analogous to definitions given in Table 3.2.1, we can define the relative Green strain measures as

$$\mathbf{E}_t = \frac{1}{2}\left(\mathbf{C}_t - \mathbf{I}\right) \qquad (3.2.61)$$

$$\mathbf{E}_t = \frac{1}{2}\left(\mathbf{I} - \mathbf{B}_t\right)$$

The relative Green strain defined in Eqn. (3.2.61) is used in various models of viscoelastic behaviour, as will be discussed later in Chapter 4.

We have seen in Sections 3.2.2 – 3.2.5, a definition of deformation gradient was given using reference (time independent) configuration. In Section 3.2.6, a relative deformation gradient, which is defined with respect to the current configuration, was defined. Depending on the material behaviour under consideration, either of these definitions and all subsequent kinematic measures can be used.

3.2.7 Time Derivatives Viewed from Different Coordinates

The time derivatives defined in Eqns. (3.2.40) and (3.2.41) are normally understood to be evaluated with respect to a non-moving (or fixed) frame. However, there are physical situations, such as description of the time derivatives such as velocity and acceleration of a particle in a spinning top. We know that the spinning top rotates about its axis, while the axis itself could wobble about a non-moving vertical axis. Therefore, the time derivatives can be described using two frames. One of the frames, which is fixed, will give a *global* description of time derivatives. The other reference frame, which is spinning along with the top, could be used to describe the *local* time derivatives of the top. The two descriptions are related to each other through the kinematics of the local (moving) frame with respect to the global (non-moving) frame. In this section, we will outline general relationships of mathematical quantities which are described in the above two frames. These relations are necessary to solve boundary and initial value problems, which are invariably solved using non-moving frames.

For modelling of engineering materials, the two most important descriptions are *co-rotational* and *convected*. The co-rotational description arises from the evaluation of the variables with respect to a rotating frame, while convected description arises from the evaluation with respect to a deforming and rotating (*i.e.* convecting) frame. Derivatives in these frames will be described in the following sections.

3.2.7.1 Co-rotational Derivatives

As mentioned above, the co-rotational derivatives involve a description with respect to a rotating frame. In this section, we will develop the relations that connect the time derivatives measured with respect to such rotating frames and a fixed frame.

Consider a non-moving reference frame that is defined using unit vectors $\mathbf{e}_i$'s and a rotating reference frame that is defined using unit vectors $\mathbf{f}_i$'s. It can be shown that $\mathbf{f}_i$'s can be obtained through an orthogonal mapping of $\mathbf{e}_i$'s using an orthogonal tensor $\mathbf{Q}(t)$ by the relation

$$\mathbf{f}_i(t) = \mathbf{Q}(t)\mathbf{e}_i. \tag{3.2.62}$$

The rate of change of the vectors $\mathbf{f}_i$ can be evaluated from Eqn. (3.2.62) using the following expression:

$$\dot{\mathbf{f}}_i = \dot{\mathbf{Q}}\mathbf{e}_i = \dot{\mathbf{Q}}\mathbf{Q}^\mathrm{T}\mathbf{f}_i = \Omega(t)\mathbf{f}_i \tag{3.2.63}$$

In deriving Eqn. (3.2.63), we have used the notation $\Omega = \dot{\mathbf{Q}}\mathbf{Q}^\mathrm{T}$, where Ω is the skew-symmetric spin tensor, which was introduced earlier in Eqn. (3.2.52). The development of the time derivatives for vector and tensor fields will be shown below.

Co-rotational Derivative of a Vector Field

Consider position vector $\mathbf{x}$ in the deformed space. We will represent $\mathbf{x}$ in terms of its components x_i in the rotating frame. Hence, the components x_i can be obtained by taking a dot product of the vector $\mathbf{x}$ with the directions $\mathbf{f}_i$. Hence, we have

$$x_i = \mathbf{x} \cdot \mathbf{f}_i \tag{3.2.64}$$

In a similar fashion, the components of the velocity associated with the time rate of change of $\mathbf{x}$ (denoted as $\overset{\circ}{\mathbf{x}}$) can be defined in terms of components $\dot{x}_i$:

$$\dot{x}_i = \overset{\circ}{\mathbf{x}} \cdot \mathbf{f}_i \tag{3.2.65}$$

The velocity components $\dot{x}_i$ can also be derived from Eqn. (3.2.64) by differentiating both sides of the equation. Hence, we get

$$\overset{\circ}{\mathbf{x}} = \dot{\mathbf{x}} \cdot \mathbf{f}_i + \mathbf{x} \cdot \dot{\mathbf{f}}_i \qquad (3.2.66)$$

Substituting for $\mathbf{f}_i$ from Eqn. (3.2.63) into Eqn. (3.2.66) and equating with the expression given in Eqn. (3.2.65), we get

$$\overset{\circ}{\mathbf{x}} = \dot{\mathbf{x}} + \boldsymbol{\Omega}^{\mathrm{T}} \mathbf{x} \qquad (3.2.67)$$

Even though Eqn. (3.2.67) has been derived for a position vector $\mathbf{x}$ in the material space, the relationship expressed in this equation can be shown to hold for any generalized vector $\mathbf{a}$, so that we have

$$\overset{\circ}{\mathbf{a}} = \dot{\mathbf{a}} + \boldsymbol{\Omega}^{\mathrm{T}} \mathbf{a} . \qquad (3.2.68)$$

$\overset{\circ}{\mathbf{a}}$, that is defined in Eqn. (3.2.68) is called as the *co-rotational derivative* of $\mathbf{a}$ and represents the rate of change of $\mathbf{a}$ with respect to the frame which itself is spinning at the rate of $\boldsymbol{\Omega}$ with respect to the fixed frame of reference. The local description that is implied by the co-rotational derivative is significant in mechanics, since it represents a convenient way the motion can be described for any vector field. However, the local description that is implied by a co-rotational derivative is limited, since it cannot be related (through any fundamental laws of motion) to other time derivatives, or to other physical quantities that are defined within the body. The total derivative that will help us in such global interactions is represented by $\dot{\mathbf{a}}$. Equation (3.2.68) is the significant relation that helps us in such interactions, by transforming a local co-rotational derivative to a global total derivative. The transformations that are represented by this equation are useful later in describing the material response, through relationships called as *constitutive relations*, that are expected to be globally defined and preserving their *material frame indifference*, with respect to any local transformations. This aspect will be described in more detail in Section 3.4, on *material frame indifference*, where these aspects are discussed in more detail.

Co-rotational Derivative of a Tensor Field

Consider a tensor field $\mathbf{T}$ and a rotating frame that is chosen to represent $\mathbf{T}$. The components of a tensor field T_{ij} can be extracted from $\mathbf{T}$ using unit vectors $\mathbf{f}_i$ s in the following way:

$$T_{ij} = \mathbf{T}\mathbf{f}_j \cdot \mathbf{f}_i \qquad (3.2.69)$$

In a similar fashion, the components of the time derivatives of $\mathbf{T}$ (designated as $\dot{T}_{ij}$), can be extracted from the co-rotational derivative of $\mathbf{T}$ (designated as $\overset{o}{\mathbf{T}}$), using the expression

$$\dot{T}_{ij} = \overset{o}{\mathbf{T}} \mathbf{f}_j \cdot \mathbf{f}_i \qquad (3.2.70)$$

We note that can also be obtained by a direct differentiation of Eqn. (3.2.69) and can be expressed in terms of the spin tensor $\boldsymbol{\Omega}$ using Eqn. (3.2.63). Hence, we get

$$\dot{T}_{ij} = \overset{o}{\mathbf{T}} \mathbf{f}_j \cdot \mathbf{f}_i = \dot{\mathbf{T}} \mathbf{f}_j \cdot \mathbf{f}_i + \mathbf{T} \dot{\mathbf{f}}_j \cdot \mathbf{f}_i + \mathbf{T} \mathbf{f}_j \cdot \dot{\mathbf{f}}_i$$

$$= \dot{\mathbf{T}} \mathbf{f}_j \cdot \mathbf{f}_i + \mathbf{T} \boldsymbol{\Omega} \mathbf{f}_j \cdot \mathbf{f}_i + \mathbf{T} \mathbf{f}_j \cdot \boldsymbol{\Omega} \mathbf{f}_i . \qquad (3.2.71)$$

Rearranging terms in Eqn. (3.2.71), we get

$$\overset{o}{\mathbf{T}} = \dot{\mathbf{T}} + \mathbf{T}\boldsymbol{\Omega} + \boldsymbol{\Omega}^{\mathrm{T}}\mathbf{T} . \qquad (3.2.72)$$

Equation (3.2.72) gives us a relation between a co-rotational time derivative of a tensor $\mathbf{T}$, *i.e.*, a derivative taken with respect to a moving reference frame rotating with an angular spin rate $\boldsymbol{\Omega}$ and the total time derivative (*i.e.*, a time derivative with respect to a fixed reference). The relationships that are represented in Eqn. (3.2.72) are significant, since they help us to relate physical quantities that are easily measured (such as co-rotational derivatives), with physical quantities that are necessary to establish global relationships. In practice, researchers have found it more convenient to use a derivative that is obtained by attaching the moving coordinate frame to be fixed to the body. In such a case, the body spin tensor $\mathbf{W}$ (introduced in Eqn. (3.2.46)) will be the same moving frame spin tensor $\boldsymbol{\Omega}$. We denote the co-rotational derivative associated with such description as $\overset{\square}{\mathbf{T}}$ and this derivative is popularly known as the *Jaumann derivative*. Thus, the Jaumann derivative is given by

$$\overset{\square}{\mathbf{T}} = \dot{\mathbf{T}} + \mathbf{T}\mathbf{W} + \mathbf{W}^{\mathrm{T}}\mathbf{T} . \qquad (3.2.73)$$

The Jaumann derivatives of a tensor and the co-rotational derivative of a vector are widely used to describe flow of fluids and flow like behaviour of solids (plasticity) as will be seen more elaborately in the subsequent chapters.

3.2.7.2 Convected Derivatives

As mentioned earlier (Section 3.2.7), *convected* description arises from the evaluation of variables with respect to a deforming and rotating (*i.e.*, *convecting*) frames. The convected co-ordinates, for describing the kinematics, are discussed in Section 2.2 and Appendix. Derivatives with respect to time in these convected coordinates are called convected derivatives and are developed below.

Let $\mathbf{x}^\tau$ be the position vector of a material point at arbitrary time τ in the fixed frame. It may be recalled that at the present time ($\tau = t$), $\mathbf{x}^\tau = \mathbf{x}^t = \mathbf{x}$. Hence, $\mathbf{x}^\tau$ can be written in terms of position vector in the convected frame $\mathbf{y}$ *viz.*

$$\mathbf{x}^\tau = \mathbf{x}^\tau(\mathbf{y}, \tau). \tag{3.2.74}$$

The position vector $\mathbf{y}$ is normally written in terms of convected co-ordinates y_i and the convected base vectors $\mathbf{g}_i$ (covariant base vectors) or $\mathbf{g}^i$ (contravariant base vectors).

$$\mathbf{y} = y^i \mathbf{g}_i = y_i \mathbf{g}^i \tag{3.2.75}$$

From the definitions of convected base vectors and Eqn. (3.2.55), we get

$$\mathbf{g}_i = \frac{\partial \mathbf{x}^\tau}{\partial y^i} = \frac{\partial \mathbf{x}^\tau}{\partial x_j} \frac{\partial x_j}{\partial y^i} = \mathbf{F}_t \mathbf{e}_i \tag{3.2.76}$$

At the present time ($\tau = t$),

$$\mathbf{F}_t = \mathbf{I} \text{ and } \mathbf{g}_i = \mathbf{e}_i \tag{3.2.77}$$

Equation (3.2.76) shows that the base vectors can be evaluated based on the knowledge of the deformation gradient and *vice versa*. It is expected that the rate of change of base vectors will incorporate information similar to the rate of change of the deformation gradient. The evaluation of the derivative with respect to τ is important since it gives us a local time derivative of any physical quantity. This local time derivative signifies a local change of the physical quantity and convected coordinates are a convenient framework to evaluate the derivative. Let us evaluate the rate of change of $\mathbf{g}_i$ at arbitrary time τ. Based on the definition of base vectors given in Eqn. (3.2.76), we get

$$\frac{\partial \mathbf{g}_i}{\partial \tau} = \frac{\partial}{\partial \tau}\left(\frac{\partial x_j^\tau \mathbf{e}_j}{\partial y_i} \right) \tag{3.2.78}$$

Since the convected coordinates y_i are independent of time, we can write

$$\frac{\partial \mathbf{g}_i}{\partial \tau} = \frac{\partial}{\partial y^i}\left(\frac{\partial x_j^\tau \mathbf{e}_j}{\partial \tau} \right) = \frac{\partial \mathbf{v}^\tau}{\partial y^i}. \tag{3.2.79}$$

where $\mathbf{v}^\tau$ is the velocity of a material point at time τ. We can use chain rule to incorporate derivative of velocity with respect to position in the fixed frame as

$$\frac{\partial \mathbf{g}_i}{\partial \tau} = \frac{\partial \mathbf{v}^\tau}{\partial y^i} = \frac{\partial \mathbf{v}^\tau}{\partial x_j^\tau} \frac{\partial x_j^\tau}{\partial y_i}. \tag{3.2.80}$$

Using index notation, we can show that Eqn. (3.2.80) can be written as

$$\frac{\partial \mathbf{g}_i}{\partial \tau} = \frac{\partial x_m^\tau \mathbf{e}_m}{\partial y_i} \left[\frac{\partial v^\tau}{\partial x_j^\tau} \mathbf{e}_j \right]^T = \mathbf{g}_i \left[\mathbf{L}(\tau) \right]^T \qquad (3.2.81)$$

It should be noted that $\mathbf{L}$ is evaluated at time τ in the above relation. At the present time,

$$\left. \frac{\partial \mathbf{g}_i}{\partial \tau} \right|_t = \mathbf{e}_i \left[\mathbf{L}(t) \right]^T \qquad (3.2.82)$$

The right-hand side of Eqn. (3.2.82) is in terms of quantities evaluated in the fixed frame. Therefore, the local derivative evaluated at time t in terms of global quantities. This derivative will be useful in evaluating convected derivatives of other mechanical variables.

Similar to Eqn. (3.2.82), we can show that

$$\frac{\partial \mathbf{g}^i}{\partial \tau} = -\mathbf{L}(\tau)\mathbf{g}^i,$$

$$\frac{\partial \mathbf{F}_t(\tau)}{\partial \tau} = \mathbf{L}(\tau)\mathbf{F}_t(\tau) \qquad (3.2.83)$$

we get the expected result given in Eqn. (3.2.59).

We can also evaluate derivatives of different strain measures such as $\mathbf{E}_t$ (Eqn. (3.2.61)).

$$\frac{\partial \mathbf{E}_t}{\partial \tau} = \mathbf{F}_t^T \underbrace{\left[\frac{1}{2} \left(\operatorname{grad} v^T + \operatorname{grad} v \right) \right]}_{\mathbf{D}(\tau)} \mathbf{F}_t \qquad (3.2.84)$$

At the present time, the above relation reduces to

$$\left. \frac{\partial \mathbf{E}_t}{\partial \tau} \right|_t = \mathbf{D}(t) \qquad (3.2.85)$$

Equation (3.2.85) signifies that the symmetric part of velocity gradient is the time rate of the relative Green Lagrange Strain tensor, evaluated at the present time. This relation will be useful when we start defining the non-linear viscoelastic models in Chapter 5.

The local derivatives $\mathbf{D}$ and other mechanical variables will be useful in the development of constitutive relations (discussed in detail in Chapter 4). It can be shown that second-order derivative of $\mathbf{E}_t$ evaluated at the current time has the following relation:

$$\left[\frac{\partial}{\partial \tau}\left(\frac{\partial \mathbf{E}_t}{\partial \tau}\right)\right]_{t} = \left\{\frac{\partial \mathbf{D}}{\partial t} + \mathbf{v}\left(\operatorname{grad}\mathbf{D}\right) + \left(\operatorname{grad}\mathbf{v}\right)^{T}\mathbf{D} + \mathbf{D}\left(\operatorname{grad}\mathbf{v}\right)\right\} \quad (3.2.86)$$

Based on the above result, we define a lower convected or covariant derivative of $\mathbf{D}$ as follows:

$$\overset{\Delta}{\mathbf{D}} = \left\{\frac{\partial \mathbf{D}}{\partial t} + \mathbf{v}\left(\operatorname{grad}\mathbf{D}\right) + \left(\operatorname{grad}\mathbf{v}\right)^{T}\mathbf{D} + \mathbf{D}\left(\operatorname{grad}\mathbf{v}\right)\right\} \quad (3.2.87)$$

Similarly, we can define the lower convected or covariant derivative $\overset{\Delta}{\mathbf{T}}$ and upper convected or contravariant derivative $\overset{\Delta}{\mathbf{T}}$ for any mechanical variable $\mathbf{T}$ as

$$\overset{\Delta}{\mathbf{T}} = \left\{\frac{\partial \mathbf{T}}{\partial t} + \mathbf{v}\left(\operatorname{grad}\mathbf{T}\right) + \mathbf{L}^{T}\mathbf{T} + \mathbf{T}\mathbf{L}\right\}$$

$$\overset{\nabla}{\mathbf{T}} = \left\{\frac{\partial \mathbf{T}}{\partial t} + \mathbf{v}\left(\operatorname{grad}\mathbf{T}\right) - \mathbf{L}\mathbf{T} - \mathbf{T}\mathbf{L}^{T}\right\}. \quad (3.2.88)$$

Combinations of these two derivatives can also be defined, such as

$$\overset{\Box}{\mathbf{T}} = \frac{1}{2}\left(\overset{\Delta}{\mathbf{T}} + \overset{\nabla}{\mathbf{T}}\right) = \frac{\partial \mathbf{T}}{\partial t} + \mathbf{v}\left(\operatorname{grad}\mathbf{T}\right) + \mathbf{W}^{T}\mathbf{T} + \mathbf{T}\mathbf{W}. \quad (3.2.89)$$

This is the same as the Jaumann derivative discussed in Eqn. (3.2.73).

In Section 3.2 on kinematics, we have defined the important mechanical variables and their appropriate spatial and time derivatives that will be useful to capture the motion of any particle as it undergoes arbitrary deformations. It is now important to use these kinematical quantities to mathematically describe the physical laws that govern the motion of any material particle. This will be outlined in Section 3.3 on balance laws, which is described next.

3.3 BALANCE LAWS

The response of any material to external stimuli is normally in accordance with fundamental laws of physics that govern its response. The physical laws are in the form of balance of mass, momentum and energy, which can be written only over a region of interest D. This region should be an order of magnitude larger than the representative volume element discussed in Chapter 1 (Section 1.1.3.2). We note that the physical laws are normally stated in the form of integral statements over the sub-region D within the material space B. However, it is possible for us to reformulate these integral statements in the form of differential statements which are valid for any material point in the material. These differential statements are commonly used to solve initial and boundary value problems in engineering analysis. In the current section on balance laws, we will be developing each of the physical laws, from their integral statements to their differential forms. To facilitate this development, we need the help of a transport theorem, which will be discussed next in Section 3.3.1.

3.3.1 Transport Theorem

In the Section 3.2, we looked at the motions of material particles, line elements and volume elements. We are usually interested in the time rate of change of integrals of variables that are defined within a sub-region D. These variables may be scalar variables like density or vector variables like momentum and forces. An integral of a scalar variable like density represents the mass of the body within the region of interest. While formulating the mass balance, we would be interested in the rate of change of the integral of this scalar variable. We note that D is a sub-region with moving boundaries. For example, the sub-region D will occupy space D^r in reference configuration, while it will occupy space D^t in the current configuration, as shown in Fig. 3.9. While evaluating the time derivative of an integral quantity with time-dependent limits, it may be useful to shift the time derivative inside the integral sign. This shift of the time derivative is facilitated by the transport theorem. The theorem is called as a transport theorem, since it relates to the transport (or motion) of integral quantities within the material space. The transport theorems in different forms are referred to as Leibnitz rule and Reynolds transport theorem. The transport theorem applied to scalar field variables will be derived below.

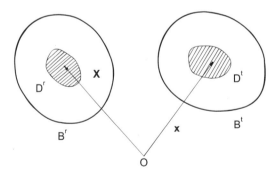

Fig. 3.9 Definition sketch for the Reynolds transport theorem

Consider a sub-region D^r within the material space B^r, of a material in the undeformed configuration. This region becomes D^t in the deformed material space B^t. Consider a scalar field variable $\phi(\mathbf{x}, t)$ (such as density or specific internal energy), that is defined in the deformed configuration and is constant within an infinitesimal volume dV_x. The transport theorem relates to the time derivative of the integral of the scalar field variable $\phi(\mathbf{x}, t)$ over the region D^t as defined below:

$$I = \frac{d}{dt} \int_{D^t} \phi(\mathbf{x}, t) dV_x$$

(3.3.1)

We recognize that the volume D^t changes with time and hence it is difficult to evaluate the integral I in the form given in Eqn. (3.3.1) above. In the following discussion, we demonstrate an example of how the integral can be evaluated. In order to facilitate the evaluation of the integral, we transform D^t from the deformed configuration to the sub-region D^r in the undeformed configuration, and then evaluate the integral. Hence, we have

$$I = \frac{d}{dt}\int_{D^t} \phi(\mathbf{x},t)dV_x = \frac{d}{dt}\int_{D^r} \phi J \, dV_X = \int_{D^r}\frac{d}{dt}(\phi J)dV_X$$

$$= \int_{D^r}\left(\dot{\phi}J + \phi\dot{J}\right)dV_X = \int_{D^t}\left(\dot{\phi}J + \phi\dot{J}\right)\frac{dV_x}{J}$$

$$= \int_{D^t}\left(\dot{\phi} + \phi \operatorname{div}\mathbf{v}\right)dV_x$$

(3.3.2)

Equation (3.3.2) is the statement of transport theorem and gives us a way of relating the time rate of change of a scalar quantity that is defined within a sub-region D^t. This expression will be used later to develop the differential forms of the balance laws. Similar statements of transport can also be derived for vector and tensor fields. For example, it is possible to prove that (Exercise Problem 3.2) given a vector field φ

$$\frac{d}{dt}\int_{D^t}\varphi \, dV_x = \int_{D^t}\left(\dot{\varphi} + \varphi \operatorname{div}\mathbf{v}\right)dV_x$$

(3.3.3)

Using Eqns. (3.3.2) and (3.3.3), we can develop the statements of balance of mass, momentum and energy. This development will be shown in the next few sections.

3.3.2 Balance of Mass

We begin our discussion with an assumption that every material is associated with a scalar field called *density* ρ, associated with each material point. The density of a material represents the inertia to motion of the material and is normally defined as the mass per unit volume of the material. The density can be expressed as a function of the current spatial co-ordinate $\mathbf{x}$ and time t. We recognize that for a sub-region D^t in the deformed configuration, the mass m_{D^t} is given by,

$$m_{D^t}(\mathbf{x},t) = \int_{D^t}\rho(\mathbf{x},t)dV_x$$

(3.3.4)

The law of conservation of mass states that matter can neither be generated nor be destroyed within a defined region D^t. This means that there are no processes which will convert a given mass into energy or *vice versa*. Therefore, the time rate of change is equal to zero. Hence, the statement of balance of mass can be written as

$$\frac{d}{dt} m_{D^t}(\mathbf{x}, t) = \frac{d}{dt} \int_{D^t} \rho(\mathbf{x}, t) \, dV_x = 0 \qquad (3.3.5)$$

Using the transport theorem of Eqn. (3.3.2), it is possible for us to reverse the order of differentiation and integration in Eqn. (3.3.5). Hence, we get

$$\frac{d}{dt} \int_{D^t} \rho(\mathbf{x}, t) \, dV_x = \int_{D^t} (\dot{\rho} + \rho \, \text{div } \mathbf{v}) dV_x = 0. \qquad (3.3.6)$$

If the integrand of Eqn. (3.3.6) is a continuous function, then Eqn. (3.3.6) will hold true for all volumes of D^t within B^t. Since the integral is zero for any arbitrary D^t, the integrand must be zero at all material points. Therefore, we get the following differential statement of the balance of mass

$$\dot{\rho} + \rho \, \text{div } \mathbf{v} = 0 = 0 \text{ for all } \mathbf{x} \in B^t. \qquad (3.3.7)$$

Equation (3.3.7) is a statement of the balance of mass within a body in the deformed configuration. It states that in a continuous system, the spatial gradients of velocity will be associated with temporal changes in the density of the system. For incompressible materials, Eqn. (3.3.7) reduces to a simple statement that

$$\text{div } \mathbf{v} = 0. \qquad (3.3.8)$$

Generally, Eqn. (3.3.8) is used with most fluids and incompressible solids such as rubbers, while Eqn. (3.3.7) is used with compressible solids and fluids.

3.3.3 Balance of Linear Momentum

We recall that Newton's second law for a particle defines a concept called force which is linked to the motion of the particle through the rate of change of momentum. These concepts can also be extended to a material body. We assume that a body $\mathcal{B}$ is being subjected to two types of forces: (a) body forces $\mathbf{b}$ acting per unit mass of the body, and (b) surface forces $\mathbf{t}$ acting per unit surface area of the body. Consider representation B^t in the deformed configuration of this body. The total force $\mathbf{f}_t$ acting on region D^t due to these distributed forces will be

$$\mathbf{f}_t = \int_{D^t} \rho \mathbf{b} \, dV_x + \int_{S^t} \mathbf{t} \, dA_x \qquad (3.3.9)$$

where S^t is the surface bounding the region D^t. Since the different particles of the body have a velocity $\mathbf{v}(\mathbf{x}, t)$ and are associated with a density ρ, it is possible to define the linear momentum $\mathbf{H_L}$ associated with the region D^t as

$$\mathbf{H_L} = \int_{D^t} \rho \mathbf{v} \, dV_x. \qquad (3.3.10)$$

Applying Newton's second law of motion to the region bounded by D^t, we can state that the external force $\mathbf{f}_t$ is equal to the rate of change of linear momentum. Hence, we can establish the following relationship:

$$\int_{D^t} \rho \mathbf{b}\, dV_x + \int_{S^t} \mathbf{t}\, dA_x = \frac{d}{dt} \int_{D^t} \rho \mathbf{v}\, dV_x. \qquad (3.3.11)$$

Equation (3.3.11) is a statement of *Linear Momentum Principle* applied to a region in a continuous body. We could use the transport theorem for vector quantities defined in Eqn. (3.3.3) and the balance of mass statement (Eqn. (3.3.7)), we can modify the time derivative of linear momentum (RHS of the Eqn. (3.3.11)) as follows:

$$\frac{d}{dt} \int_{D^t} \rho \mathbf{v}\, dV_x = \int_{D^t} \left\{ \left(\dot{\overline{\rho \mathbf{v}}} \right) + \rho \mathbf{v}\operatorname{div} \mathbf{v} \right\} dV_x \qquad (3.3.12)$$

$$= \int_{D^t} \left\{ \rho \dot{\mathbf{v}} + \mathbf{v}\left(\dot{\rho} + \rho \operatorname{div} \mathbf{v} \right) \right\} dV_x = \int_{D^t} \rho \dot{\mathbf{v}}\, dV_x.$$

Therefore, Eqn. (3.3.11) reduces to

$$\int_{D^t} \rho \mathbf{b}\, dV_x + \int_{S^t} \mathbf{t}\, dA_x = \int_{D^t} \rho \dot{\mathbf{v}}\, dV_x. \qquad (3.3.13)$$

The L.H.S. of Eqn. (3.3.13) has the surface force $\mathbf{t}$, which is also referred to as the traction force. We note that this traction force $\mathbf{t}$ depends upon the orientation of the surface with unit normal vector $\mathbf{n}$ at any given point $\mathbf{x}$. Hence, we can write

$$\mathbf{t} = \mathbf{t}(\mathbf{x}, \mathbf{n}, t). \qquad (3.3.14)$$

We further note that the $\mathbf{t}$ is a force that is self-equilibrating for any internal point $\mathbf{x}$. This means that it has following properties:

$$\mathbf{t}(\mathbf{x}, \mathbf{n}, t) = -\mathbf{t}(\mathbf{x}, -\mathbf{n}, t). \qquad (3.3.15)$$

Figure 3.10 shows an illustration of Eqn. (3.3.15). In this figure, we consider a configuration of the body $\mathcal{B}$ being split into two halves by an imaginary plane. The traction force $\mathbf{t}$ acting on region 1 (on the dividing plane represented by unit normal vector $\mathbf{n}$) represents the effect of region 2 at the dividing plane. Similarly, there will be a traction force by region 1 on region 2 (on the dividing plane represented by unit normal vector $-\mathbf{n}$).

Equation (3.3.15) states that the two traction forces will be equal in magnitude and opposite in direction.

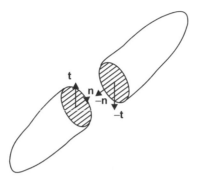

Fig. 3.10 Definition of traction

Further, it is possible to show that the traction force **t** can be related to a tensor $\sigma(\mathbf{x},\ t)$ as follows,

$$\mathbf{t}(\mathbf{x},\ \mathbf{n},\ t) = \sigma^T(\mathbf{x},\ t)\mathbf{n}. \tag{3.3.16}$$

The equation stated in Eqn. (3.3.16) tells us that it is possible to extract the area dependence of the traction force, by defining a tensor field $\sigma(\mathbf{x},\ t)$, which will depend only on the spatial and temporal co-ordinates. The tensor σ is called as the *Cauchy stress tensor*. The stress tensor has the same units as the traction t (force per unit surface area in the deformed configuration) and represents the intensity of load that any unit area in the deformed configuration experiences in a material.

Using index notation, σ_{ij} can be expressed as

$$\sigma_{ij} = \sigma\mathbf{e}_j \cdot \mathbf{e}_i = \mathbf{e}_j \cdot \sigma^T\mathbf{e}_i = \mathbf{e}_j \cdot \mathbf{t}\,(\mathbf{x},\ \mathbf{e}_i,\ t) = t_j\,(\mathbf{x},\ \mathbf{e}_i,\ t). \tag{3.3.17}$$

Equation (3.3.17) helps us to interpret σ_{ij} to be the traction force acting in the direction $\mathbf{e}_j$ per unit surface area, whose normal is oriented along the direction $\mathbf{e}_i$ in the deformed configuration. The components of σ on a typical plane normal to $\mathbf{e}_1$ are shown in Fig. 3.11. It is important to note that the Cauchy stress in Eqn. (3.3.16) is defined with respect to the current configuration. We will find that this definition is a useful concept to express the state of stress. However, on the boundaries, it is often not a convenient concept for solving the boundary value problems, since the exact shape of the boundary will not be known *a priori*. This is illustrated in the stretch of an elastic membrane having a circular boundary, subject to a uniform tension, as shown in Fig. 3.12. We find that even though the initial geometry of the inner cut-out is defined to be a circle, the deformed configuration of the circle is not known *a priori*. We do not know **n** on the deformed circle. The traction on the deformed circle is known always to be zero, as the cut-out continues to be stress-free. Therefore, Cauchy stresses cannot be evaluated at the deformed circle. These concepts will be discussed more in detail in later sections.

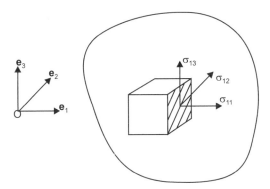

Fig. 3.11 Components of Cauchy stress tensor

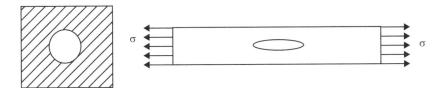

Fig. 3.12 Deformations of sheet with a cut-out. (a) Undeformed sheet with a
circular cut-out. (b) Deformed sheet due to a uniform tension, showing an
elliptical cut-out.

We return to the statement of balance of linear momentum given in Eqn.
(3.3.13). We substitute Eqn. (3.3.16) into Eqn. (3.3.13) to obtain

$$\int_{D^t} \rho \mathbf{b}\, dV_x + \int_{S^t} \boldsymbol{\sigma}^T(\mathbf{x},t)\, \mathbf{n}\, dA_x \;=\; \int_{D^t} \rho \dot{\mathbf{v}}\, dV_x. \qquad (3.3.18)$$

We can further use Gauss's divergence theorem (Eqn. (2.4.1)) to convert the
surface integral on the LHS of Eqn. (3.3.18) into a volume integral to obtain

$$\int_{D^t} \rho \mathbf{b}\, dV_x + \int_{D^t} \operatorname{div} \boldsymbol{\sigma}^T(\mathbf{x},t)\, dV_x \;=\; \int_{D^t} \rho \dot{\mathbf{v}}\, dV_x. \qquad (3.3.19)$$

Since Eqn. (3.3.19) must hold true for every region D^t of the body, (and
recognising that $\dot{\mathbf{v}} = \mathbf{a}$), we can arrive at the point wise differential statement
of the balance of linear momentum to be,

$$\operatorname{div} \boldsymbol{\sigma}^T + \rho \mathbf{b} = \rho \mathbf{a}. \qquad (3.3.20)$$

This equation is also referred to as the *Linear Momentum Principle (LMP)*. It
is also called *Equation of Motion*.

At a component level, Eqn. (3.3.20) can be written as

$$\frac{\partial \sigma_{ji}}{\partial x_i} + \rho b_i = \rho a_i. \qquad (3.3.21)$$

The balance of linear momentum is normally referred to as *stress-equilibrium equations* in elasticity theory and the stress-equilibrium equations have the same form as Eqn. (3.3.21). However, there are differences between the derivations that are made in elasticity books and the more general statements that we are trying to make here. In elasticity theory, which is a theory of infinitesimal deformations, the distinction between the deformed and undeformed configurations is not recognized. Therefore, the following differences may be noted between the current formulation and the elasticity theory:

(a) The stresses defined in elasticity theory are nominal stresses that are defined with respect to the undeformed configuration. However, the stresses defined in Eqn. (3.3.21) are the Cauchy stresses that are specifically defined in the deformed configuration.

(b) The spatial derivatives that are mentioned in Eqn. (3.3.21) are with respect to the current configuration, while the spatial derivatives that are normally defined in the stress equilibrium equations are with respect to the undeformed configuration.

The stress equilibrium equations can be considered as a specific form of the more general statement of the balance of linear momentum that is stated in Eqn. (3.3.20).

3.3.4 Balance of Angular Momentum

We will now consider the moment caused by the forces $\mathbf{b}$ acting on the volume element D^t and $\mathbf{t}$ acting on the surface S^t about any arbitrary point O which is at a distance $\mathbf{x}$ from the volume element. The total moment $\mathbf{M}_t$ induced by these forces about the point O is given by

$$\mathbf{M}_t = \int_{D^t} \mathbf{x} \times \rho \mathbf{b}\, dV_x + \int_{S^t} \mathbf{x} \times \mathbf{t}\, dA_x. \qquad (3.3.22)$$

The induced moment $\mathbf{M}_t$ will cause rotations of the various particles in the body about the point O, so that it will be possible for us to define an angular momentum $\mathbf{H}_A$ associated with these rotations, which is defined as

$$\mathbf{H}_A = \int_{D^t} \mathbf{x} \times \rho \mathbf{v}\, dV_x. \qquad (3.3.23)$$

Once again, applying Newton's second law of motion to the rotations induced in the region bounded by D^t, we can state that the external moment $\mathbf{M}_t$ is equal to the rate of change of angular momentum $\dot{\mathbf{H}}_A$. Hence, we can establish the following relationship:

$$\int_{D^t} \mathbf{x} \times \rho \mathbf{b}\, dV_x + \int_{S^t} \mathbf{x} \times \mathbf{t}\, dA_x = \frac{d}{dt} \int_{D^t} \mathbf{x} \times \rho \mathbf{v}\, dV_x = \int_{D^t} \mathbf{x} \times \rho \mathbf{a}\, dV_x. \qquad (3.3.24)$$

Equation (3.3.24) is a statement of the balance of angular momentum or *Angular Momentum Principle,* applied to a small region in a body. In a procedure analogous to the derivation of the balance of linear momentum, it is possible for us to convert the surface tractions **t** into Cauchy stress tensor σ. We can also apply the Gauss divergence theorem to convert the surface integral in Eqn. (3.3.24) into a volume integral. As a consequence of all these transformations, it can be shown that

$$\sigma = \sigma^{T}. \tag{3.3.25}$$

Equation (3.3.25) tells us that the Cauchy stress tensor is symmetric, as a consequence of the balance of angular momentum. It is important to note that this property is unique only to Cauchy stresses, which are defined with respect to the deformed configurations. We will later see that not all stress measures are symmetric. Further, we note that the stresses are implicitly proven to be symmetric in infinitesimal deformation theory, which is used in subjects like Theory of Elasticity, again as a consequence of the Angular Momentum Principle, which is applied to the deformed configuration. However, since there is no distinction between deformed and undeformed configurations in infinitesimal theory, it is assumed there that the stresses are *always* symmetric. We observe now that it is not possible to make such a statement for finite deformations for all stress measures.

3.3.5 Work Energy Identity

In addition to the statements of the balances of linear momentum (Eqn. (3.3.20)) and angular momentum (Eqn. (3.3.25)), it is possible for us to quantify the effect of external forces on the body, through statements of energy. We know, for example, that all external loads acting on the body should do some work on the deformable body and this external work is stored in the body either as kinetic energy (due to motion of the body) or as internal energy (due to internal deformations in the body). We will now get a mathematical expression relating these two quantities in a body undergoing finite deformations. It may be noted that the relation that is being derived is a useful expression that relates the quantities that are defined in the balance of linear and angular momentums.

Let us denote the rate of work on the deformable body due to external body forces **b** and surface forces **t** as $\dot{W}$. Since we know that the rate of work is a product of forces and velocities acting at elemental areas and volumes of the body, we can obtain an expression for as

$$\dot{W} = \int_{S^t} \mathbf{t} \cdot \mathbf{v} dA_x + \int_{D^t} \rho \mathbf{b} \cdot \mathbf{v} dV_x . \tag{3.3.26}$$

Equation (3.3.26) gives us an expression of the rate of work as induced by the external forces on the body. These external agents will cause a change in the energy state in the body, in the form of a change in kinetic energy or the rate of internal working. We will now attempt to derive this relationship. We will begin our derivation by writing Eqn. (3.3.26) in its component form and by converting the surface integral in Eqn. (3.3.26) into a volume integral using the Gauss divergence theorem. Hence, we have

$$\dot{W} = \int_{S^t} \mathbf{t} \cdot \mathbf{v} dA_x + \int_{D^t} \rho \mathbf{b} \cdot \mathbf{v} dV_x = \int_{S^t} t_i v_i dA_x + \int_{D^t} \rho b_i v_i dV_x \qquad (3.3.27)$$

$$= \int_{D^t} \sigma_{ij} \frac{\partial v_i}{\partial x_j} dV_x + \int_{D^t} \left(\frac{\partial \sigma_{ij}}{\partial x_j} + \rho b_i \right) v_i dV_x.$$

We note that we can now use the Linear Momentum Principle (Eqn. (3.3.20)) to substitute for the terms in the brackets of the second term of Eqn. (3.3.27) and use the definition of velocity gradient L_{ij} in the first term of Eqn. (3.3.27). Hence, we can reduce Eqn. (3.3.27) to the following:

$$\dot{W} = \int_{D^t} \sigma_{ij} L_{ij} dV_x + \int_{D^t} \frac{1}{2} \rho \left(\dot{v_i v_i} \right) dV_x$$

$$= \int_{D^t} tr(\boldsymbol{\sigma} \mathbf{L}) dV_x + \int_{D^t} \frac{1}{2} \rho \left(\dot{v_i v_i} \right) dV_x$$

$$= \int_{D^t} tr(\boldsymbol{\sigma} \mathbf{D}) dV_x + \frac{d}{dt} \int_{D^t} \frac{1}{2} \rho \left(v_i v_i \right) dV_x \qquad (3.3.28)$$

From Eqns. (3.3.27) and (3.3.28), we get the statement of work energy identity as

$$\int_{S^t} \mathbf{t} \cdot \mathbf{v} dA_x + \int_{D^t} \rho \mathbf{b} \cdot \mathbf{v} dV_x = \int_{D^t} tr(\boldsymbol{\sigma} \mathbf{D}) dV_x + \frac{d}{dt} \int_{D^t} \frac{1}{2} \rho \mathbf{v} \cdot \mathbf{v} dV_x. \qquad (3.3.29)$$

Equation (3.3.29) is called as the *work energy identity* and is an important relationship which relates the rate of work done by external agents like body forces **b** and surface tractions **t**, and their effects as experienced by individual particles of the body. We note that the first term in the right hand side of Eqn. (3.3.29) represents the rate of internal working by the stresses σ in the body. The second term on the right hand side of Eqn. (3.3.29) represents the rate of increase in the kinetic energy in the system. In other words, Eqn. (3.3.29) tells us that the only effect that external agents can have in a *conservative* (*non-dissipative*) deformation of a continuum is to increase the kinetic energy of the system as well as to increase the rate at which internal agents like

Cauchy stresses do work on the body. The implications of energy *dissipation* associated with the above statement will be discussed later in this chapter. The work energy identity will be used in statements of balance of energy later in the next section.

3.3.6 Thermodynamic Principles

In the previous section, we have discussed the mechanical agents that do work on any body, which result in a change in the mechanical energy in the body. We will now formulate the influence of non-mechanical agents in causing a mechanical response in a continuum. The non-mechanical agents that can cause changes in *energy of the body* are thermal, electrical and magnetic agents. One of the most common non-mechanical agents causing mechanical response is a thermal agent, usually referred to as *heat*. Heat is visualized as form of transferring energy from a control volume within a body to another. Usually, we learn the relations among mechanical work, heat and *non-mechanical energy* in Thermodynamics. *Internal energy, enthalpy, Gibbs free energy* and *Helmholtz free energy* are the examples of non-mechanical energies and will be discussed later. The description of the two laws of thermodynamics considering finite deformation of bodies is given in the next two sections.

3.3.6.1 First Law of Thermodynamics

We learn that the first law of thermodynamics states that internal energy of a system changes in response to the mechanical work on (by) the system and the heat transfer into (out of) the system. This concept can be translated in the form of energy balance of a material undergoing finite deformation. The expression for the energy balance will be developed in this section.

We will postulate that it will be possible to supply heat throughout the body at the rate of r per unit mass and at the rate of h per unit area across the surface of any typical volume D^t with a surface area S^t. Hence, the total rate of heating $\dot{H}_t$ in the volume element D^t is given by

$$\dot{H}_t = \int_{D^t} \rho r dV_x + \int_{S^t} h dA_x. \qquad (3.3.30)$$

We will also now postulate the existence of an internal variable called as internal energy that is defined as the energy stored inside the material per unit mass and denote it by the symbol ε. Similar to density, ε is a scalar field and can be defined as a function of current configuration and time, $\varepsilon(\mathbf{x}, t)$.

The *first law of thermodynamics* states that effect of all external agents like mechanical forces (Eqn. (3.3.26)) and heat ($\dot{W}$, Eqn. (3.3.30)) on the volume D^t bounded by S^t would be two fold; it could either (a) cause motion in the body and thereby change its kinetic energy, or (b) change the internal energy of the body. Therefore, we obtain

$$\dot{W} + \dot{H}_t = \frac{d}{dt} \int_{D^t} \frac{1}{2} \rho \mathbf{v} \cdot \mathbf{v} \, dV_x + \frac{d}{dt} \int_{D^t} \rho \varepsilon \, dV_x. \tag{3.3.31}$$

Substituting for $\dot{W}$ and $\dot{H}_t$ from Eqns. (3.3.26) and (3.3.30), respectively, in Eqn. (3.3.31), we get

$$\int_{S^t} \mathbf{t} \cdot \mathbf{v} \, dA_x + \int_{D^t} \rho \mathbf{b} \cdot \mathbf{v} \, dV_x + \int_{D^t} \rho r \, dV_x + \int_{S^t} h \, dA_x =$$

$$\frac{d}{dt} \int_{D^t} \frac{1}{2} \rho \mathbf{v} \cdot \mathbf{v} \, dV_x + \frac{d}{dt} \int_{D^t} \rho \varepsilon \, dV_x. \tag{3.3.32}$$

Substituting the work-energy identity from Eqn. (3.3.29) into Eqn. (3.3.32), we obtain the modified form of the first law of thermodynamics for a continuum as

$$\int_{D^t} \rho r \, dV_x + \int_{S^t} h \, dA_x + \int_{D^t} \boldsymbol{\sigma} : \mathbf{D} \, dV_x = \frac{d}{dt} \int_{D^t} \rho \varepsilon \, dV_x. \tag{3.3.33}$$

Similar to mass and momentum balance equations, we will now develop the energy balance in the differential form. We recognize that the rate of surface heating h is a function of the area across which heat is being supplied to the body. It is possible to decouple this dependence on the normal vector $\mathbf{n}$ by defining a heat flux vector $\mathbf{q}$ (similar to the traction and stress tensor transformation relation (Eqn. (3.3.16))) as follows:

$$h(\mathbf{x}, \mathbf{n}, t) = \mathbf{q}(\mathbf{x}, t) \cdot \mathbf{n}. \tag{3.3.34}$$

Substituting Eqn. (3.3.34) in Eqn. (3.3.35), and using the Gauss divergence theorem to convert the surface integral into a volume integral, it is possible to get the pointwise differential form of the first law of thermodynamics as

$$\rho r + \text{div} \mathbf{q} + \boldsymbol{\sigma} : \mathbf{D} = \rho \dot{\varepsilon}. \tag{3.3.35}$$

Equation (3.3.35) is an important relationship that relates the energies that are associated with any deformable system. It is significant since it introduces the concept of an internal energy which is time dependent. It is relevant even in the absence of thermal forces. For elastic systems, the internal energy is normally postulated to be the *strain energy density* in the system. The additive nature, in which the effect of external agents is expressed in Eqn. (3.3.31), makes it amenable for the inclusion of other causes of mechanical action, such as electrical potentials or magnetic fields.

3.3.6.2 Second Law of Thermodynamics

We learn that the energy balance by itself is not sufficient to describe processes that involve extraction of work from heating or *vice versa*. Let us consider a simple example of thermal energy balance. A hot stream and a cold stream are mixed to form a lukewarm stream. In terms of energy balance, we can write that

$$\text{Energy}_{\text{hot}} + \text{Energy}_{\text{cold}} = \text{Energy}_{\text{lukewarm}}.$$

This statement of energy balance does not tell us that it is easy to mix the two streams and get the lukewarm stream, but it will be *difficult* to split the lukewarm stream into one hot stream and one cold stream. In thermodynamics, we have learnt that such difficulties can be explained on the basis of a function called *Entropy*. Similar to density and internal energy, entropy is a scalar field defined for each material point. We define $\eta(\mathbf{x}, t)$ to be the entropy per unit mass and it can be defined as a function of current configuration and time. Entropy *is said to be* a measure of *disorder* and/or *dissipation* at any given point in continuum. The concept of entropy is used to address the following issues such as approach to equilibrium, the spontaneity of processes, the feasibility of certain machines/processes/ideas, the understanding of microscopic/ecological/low temperature/cosmological world and the behaviour of complex materials

In order to develop the concepts of thermodynamics further, we need to define a thermal indicator called the temperature. It is understood to distinguish between *hot* and *cold* states of a material. It is also said to be an indicator of the heat content at any given point in the materials. The temperature $\theta(\mathbf{x}, t)$, is a scalar field and can be defined to be a function of the current configuration and time.

We will now proceed to develop the second law of thermodynamics, which will be applicable to any control volume in the continuum. The second law of thermodynamics as applied to a region D^t, states that the heat generated in D^t per unit temperature, must be less than or equal to the rate of increase of entropy in D^t. Since heat can be generated in the system either by internal heat source r or through the surface inputs h, the integral form of the second law of thermodynamics will be

$$\int_{S^t} \frac{h}{\theta} dA_x + \int_{D^t} \rho \frac{r}{\theta} dV_x \leq \frac{d}{dt} \int_{D^t} \rho \eta \, dV_x \qquad (3.3.36)$$

We can substitute for surface heating h in terms of heat flux $\mathbf{q}$, using Eqn. (3.3.34) in the above equation. We can further use the Gauss divergence theorem and arrive at the differential form of the second law of thermodynamics,

$$\text{div}\left(\frac{\mathbf{q}}{\theta}\right) + \rho \frac{r}{\theta} \leq \rho \dot{\eta} \qquad (3.3.37)$$

Equation (3.3.37), which is a statement of the second law of thermodynamics as applied to a point in a continuum, is called as the *Clausius-Duhem Inequality*. This inequality, though not useful to solve for the involved variables (being an inequality), is very useful in establishing the bounds for any existing or postulated variables. The implications and the use of this inequality will be discussed in more detail in Section 3.4, when we describe constitutive relations.

Internal energy and entropy are both *thermodynamic functions*, which are very useful to describe material behaviour. In the next section, we introduce these and related thermodynamics functions.

3.3.6.3 Alternate Energy Measures in Thermodynamics

In the previous sections, both internal energy and entropy were introduced as functions of configuration and time, *i.e.* ε (**x**, t) and η (**x**, t). When we postulated the existence of internal energy for a material, we made an implicit assumption that the internal energy at a material point can be estimated. We assume that internal energy depends on several other attributes such as deformation gradient at a material point. A description of all the attributes of a material point will define the *state* associated with the material point. From thermodynamics, we know that there are three variables that are involved in the description of the state of a material point. We also know from thermodynamics that only two of these variables are independent and the third variable can be expressed in terms of the other two variables. For example, we can write

$$\varepsilon = \varepsilon(\eta, \mathbf{F}). \qquad (3.3.38)$$

In writing Eqn. (3.3.38), we have written the attributes η and **F** in place of **x** and t. In this description, space and time become implicit. In other words, as position and time change, the state variable of a material point changes. It should be noted that depending on a specific material under consideration, more number of state variables may have to be considered for the description of the state of a material point.

In Section 3.2, several measures of deformation were introduced. Similarly, Eqn. (3.3.38) can be written with different measures. However, for the discussion below we will use a kinematical variable κ_v which can be any of the strain measures such as deformation gradient, Green strain, infinitesimal strain, etc. Therefore, we can write

$$\varepsilon = \varepsilon(\eta, \kappa_v) \qquad (3.3.39)$$

An example of kinematical variable is *specific volume* (v) or the volume per unit mass (=1/ρ). To develop alternate descriptions of the state through the use of other state variables such as temperature and stress, we will consider a variation of the internal energy ε described in Eqn. (3.3.40) below:

$$d\varepsilon = \left(\frac{\partial \varepsilon}{\partial \eta}\right)_{\kappa_v} d\eta + \left(\frac{\partial \varepsilon}{\partial \kappa_v}\right)_{\eta} d\kappa_v. \quad (3.3.40)$$

From thermodynamics, we also know that

$$\left(\frac{\partial \varepsilon}{\partial \eta}\right)_{\kappa_v} = \theta. \quad (3.3.41)$$

Similar to Equation (3.3.41), we can define the derivative with respect to kinematic variable κ_v as another kinetic variable s_v which is given by

$$\left(\frac{\partial \varepsilon}{\partial \kappa_v}\right)_{\eta} = s_v \quad (3.3.42)$$

s_v can be the Cauchy stress as defined earlier (or other measures of stress that will be described in Section 3.3.7.1). Equations (3.3.41) and (3.3.42) define temperature and kinetic variable from a thermodynamic standpoint, as changes in internal energy with respect to entropy and kinematic variable respectively. The above equations will also help us to define alternate measures of energy. These alternate measures can be developed by replacing η with θ and/or κ_v with s_v. The alternate energy measures that are commonly used in the literature are listed in Table 3.2.

Table 3.2 Alternate Energy Measures

Energy measure	Symbolic description
Internal energy	$\varepsilon = \varepsilon(\eta, \kappa_v)$
Helmholtz free energy	$\psi = \psi(\theta, \kappa_v)$
Gibbs free energy	$\zeta = \zeta(\theta, s_v)$
Enthalpy	$\xi = \xi(\eta, s_v)$

Table 3.2 describes the variables based on which the energy measures are most commonly defined. In other words, changes in the variables are related to the change in energy measures. In engineering practice, among the four variables, effecting controlled changes in η is more difficult. On the other hand, it is easier to keep the temperature constant or a kinematic variable like the strain rate constant. Therefore, Gibbs free energy and Helmholtz free energy are used more often in modeling of engineering materials.

As was mentioned in Section 1.1.1, many engineering materials consist of multi-components and multiple phases. The balance laws and thermodynamic energy measures described in Sections 3.3.6.1-3.3.6.3 can be extended for such complex materials. A preliminary discussion on the extension of balance laws to multi-component and multiphase materials is given in Section 3.3.9. In the next two sections, we will examine the balance laws described so far in terms of alternate descriptions and determinacy.

3.3.7 Referential Description of Balance Laws

In this chapter, we have defined several variables to describe the material behaviour. There are kinematic quantities such as the position $\mathbf{x}$, velocity $\mathbf{v}$ and acceleration $\mathbf{a}$, all of which are defined only in the deformed configuration. Some of these variables such as density, internal energy and temperature were introduced as functions of current configuration and time. However, it is entirely possible to express these variables as functions of reference or undeformed configuration ($\mathbf{X}$). We also have specific quantities like the density ρ, the body forces $\mathbf{b}$, the rate of heating r, the internal energy ε and the entropy η, which have been defined per unit volume of the material. We further have quantities that are defined per unit area such as the surface traction $\mathbf{t}$, the Cauchy stress tensor σ and the heat flux vector $\mathbf{q}$. We can choose the deformed or the undeformed configuration to define the area and volume of an element. For example, we can choose to denote a referential description of density ρ_o (which will represent the mass per unit undeformed volume of a volume element). We will denote the referential representation of the surface traction as $\mathbf{s}$ (contact force per unit surface area in the undeformed configuration), the stress tensor as $\mathbf{S}$ (stress tensor defined on the basis of unit surface area in the undeformed configuration) and the heat flux as $\mathbf{q}_o$ (heat flux vector defined on the basis of unit surface area in the undeformed configuration).

We note that a referential description of the quantities is many times preferred because the reference configuration is known a-priori and does not change with time, while the deformed configuration is something that keeps changing and may be difficult to estimate at each time. However, many laws of physics, such as the Newton's second law, are usually written in the deformed configuration. Hence, it may be useful to convert the variables in the deformed configuration, into equivalent quantities in the undeformed configuration, so that we have the flexibility to express the laws of physics either in the deformed configuration or in the undeformed configuration. We will illustrate a few of these conversions in the following discussion.

3.3.7.1 Relations between Variables in Deformed and Undeformed Configurations

Density

Consider the total mass M enclosed in a control volume D^t (in the deformed configuration), whose volume in the undeformed configuration is D^r. We can express M in terms of the density in the deformed configuration ρ, or the density in the undeformed configuration ρ_o as shown below.

$$M = \int_{D^t} \rho dV_x = \int_{D^r} \rho_o\, dV_X = \int_{D^t} \rho_o \frac{dV_x}{J}. \tag{3.3.43}$$

Equation (3.2.11) has been used to relate dV_X to dV_x in Eqn. (3.3.43). From first and third integrals of Eqn. (3.3.43), we get

$$\rho_o = J\rho . \tag{3.3.44}$$

Stresses

Let us denote $\mathbf{S}_t$ as the force acting on an area element dA_x in the deformed configuration, due to surface traction $\mathbf{t}$ acting on the surface. Using the transformation relationship between areas in the undeformed configuration dA_x and the area in the deformed configuration dA_X given in Eqn. (3.2.16), we get

$$\mathbf{S}_t = \mathbf{t}\, dA_x = \sigma\, \mathbf{n}\, dA_x = J\sigma\, \mathbf{F}^{-T} \mathbf{n}_o\, dA_X, \tag{3.3.45}$$

where $\mathbf{n}_o$ is the unit normal vector for the area element defined in the undeformed configuration. Equation (3.3.45) suggests that it is possible to define a stress tensor $\mathbf{S}$ and a traction vector $\mathbf{s}$ in the undeformed configuration as

$$\mathbf{S} = J\sigma\, \mathbf{F}^{-T} \text{ and } \mathbf{s} = \mathbf{S}\, \mathbf{n}_o = J\sigma\, \mathbf{F}^{-T} \mathbf{n}_o. \tag{3.3.46}$$

The stress tensor $\mathbf{S}$, which is naturally defined in the undeformed configuration, is called as the 1st *Piola-Kirchhoff Stress* and the traction $\mathbf{s}$ is called as the 1st *Piola-Kirchhoff traction*. The stress $\mathbf{S}$ is also popularly known as *Nominal stress* in the literature.

The components of $\mathbf{S}$, S_{ij} need to be carefully interpreted. We note that S_{ij} refer to the force acting on the deformed body in the ith direction, over an area that is oriented along j^{th} direction in the undeformed configuration. This is in contrast the components of σ, where j^{th} direction is referred to an area element in the deformed configuration. The components S_{ij}, for a typical element, when referred to the rectangular co-ordinate frame, are shown in Fig. 3.13.

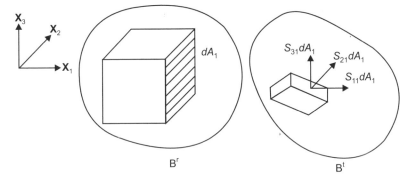

Fig. 3.13 Components of 1st Piola-Kirchhoff stress tensor (dA_1 is the area

Symmetry of S

We could invert Eqn. (3.3.46) to write σ in terms of **S**. Hence, we get

$$\sigma = \frac{1}{J}\mathbf{S}\mathbf{F}^{T}. \tag{3.3.47}$$

Further, using the symmetry relations for σ as given in Eqn. (3.3.25), we obtain the following relationship:

$$\mathbf{S}\mathbf{F}^{T} = \mathbf{F}\mathbf{S}^{T}. \tag{3.3.48}$$

Equation (3.3.48) clearly tells us that **S** is not symmetric. Hence, we note that the 1st Piola-Kirchhoff stress **S**, even though convenient to use by virtue of its reference to the undeformed areas, will give us tensors that are not symmetric. This lack of symmetry is undesirable, especially when one is doing numerical computations with these matrices, in a boundary value problem. Hence, several alternate stress measures are used in the literature, which have the dual advantage of the reference to undeformed configuration and symmetry. They are listed in Table 3.3.

Table 3.3 Different Stress Measures Used in the Literature

S. No.	Name	Expression
1	2nd Piola-Kirchhoff stress ($\mathbf{S}_1$)	$J\mathbf{F}^{-1}\sigma\mathbf{F}^{-T}$ or $\mathbf{F}^{-1}\mathbf{S}$
2	Biot stress	$\frac{1}{2}(\mathbf{S}^T\mathbf{R} + \mathbf{R}^T\mathbf{S})$

3.3.7.2 Statement of the Balance Laws in Reference Configuration

It is possible for us to restate each one of the balance laws (derived earlier) in their differential forms, in the reference configuration (see Exercise Problem 3.16). The final forms of each of these balance laws are given in Table 3.4. We note that in this statement, we will be using the Div operator, which will refer to the derivatives taken with reference configuration. Hence, we have

$$\text{Div}\,\mathbf{S} = \frac{\partial S_{ij}}{\partial X_j} \tag{3.3.49}$$

Table 3.4 Balance Laws in Reference Configuration

Balance law	Differential form of the law
Balance of mass	$\rho_o = \rho J$
Balance of linear momentum	$\text{Div}\,\mathbf{S} + \rho_o\mathbf{b} = \rho_o\dot{\mathbf{v}}$
Balance of angular momentum	$\mathbf{S}\,\mathbf{F}^T = \mathbf{F}\,\mathbf{S}^T$
First law of thermodynamics	$\mathbf{S}:\dot{\mathbf{F}} + \text{Div}\,\mathbf{q}_o + \rho_o r = \rho_o\dot{\varepsilon}$
Second law of thermodynamics	$\text{Div}\left(\dfrac{\mathbf{q}_o}{\theta}\right) + \dfrac{\rho_o r}{\theta} \leq \rho_o\dot{\eta}$

As mentioned in Section 3.3.6.3, alternate energy measures can be defined to describe the laws of thermodynamics. If we choose the kinematic variable κ_v to be the deformation gradient $\mathbf{F}$, the kinetic variable s_v is the 1st Piola-Kirchhoff stress $\mathbf{S}$. Therefore, based on Eqn. (3.3.42), we can write

$$\left(\frac{\partial \varepsilon}{\partial \mathbf{F}}\right)_\eta = \frac{1}{\rho_0}\mathbf{S} \tag{3.3.50}$$

Analogous to the energy variables described in Table 3.2, we can define the alternate energy measures in terms of deformation gradient and the 1st Piola-Kirchhoff stress

$$\psi = \varepsilon - \eta\theta, \; E = \varepsilon - \frac{1}{\rho_0}\mathbf{S:F} \quad \text{and} \quad \zeta = \varepsilon - \eta\theta - \frac{1}{\rho_o}\mathbf{S:F}. \tag{3.3.51}$$

3.3.8 Indeterminate Nature of the Balance Laws

We recall the following differential forms of the balance laws introduced in Sections 3.3.2-3.3.6 (Eqns. (3.3.7, 3.3.20, 3.3.25, 3.3.35, 3.3.37)).

$$\dot{\rho} + \rho \, \text{div} \, \mathbf{v} = 0$$
$$\text{div} \, \sigma^T + \rho\mathbf{b} = \rho\mathbf{a}$$
$$\sigma = \sigma^T$$
$$\rho r + \text{div} \, \mathbf{q} + \sigma : \mathbf{D} = \rho\dot{\varepsilon}$$

$$\text{div}\left(\frac{\mathbf{q}}{\theta}\right) + \rho\frac{r}{\theta} \leq \rho\dot{\eta} \tag{3.3.52}$$

Alternately, equivalent statements in the undeformed configuration were introduced in section 3.3.7 (Table 3.4). Let us examine the number of equations in Eqn. (3.3.52). We note that the last of these expressions is an inequality and hence cannot be used to solve for the variables appearing in the equations. Further, Eqn. (3.3.25) states that σ is symmetric and can be used to reduce the number of variables that we count in the tensor σ. Therefore, we have one equation in Eqn. (3.3.7), three equations in Eqn. (3.3.20) and one equation in Eqn. (3.3.35). Hence, we have five equations available with us.

Let us count the number of variables that we are dealing with in the description of continuum. We have the following number of variables:

(a) One variable associated with the density ρ,
(b) Six variables associated with the Cauchy stress tensor σ,
(c) Three variables associated with the velocity $\mathbf{v}$,
(d) Three variables associated with the heat flux vector $\mathbf{q}$,
(e) One variable associated with the internal energy ε,
(f) One variable associated with the temperature θ.

Thus, we find that we need a total number of 15 variables to describe the complete thermo-mechanical system.

We find that there are 5 equations and 15 variables to be determined in the system. Hence, the system is indeterminate and cannot be solved for, unless we have additional relationships that relate the involved variables. These additional relationships are normally provided for by the Constitutive Relations, which will tell us more about how the kinematic variables such as velocities are related to the non-kinematic variables such as stresses in a material. These relations will be dealt with in more detail in Section 3.4.

3.3.9 A note on Multiphase and Multi-component Materials

When multiple phases exist in the material, the balance laws described in Sections 3.3.2-3.3.6 will have to be modified. We recall that all kinematic and kinetic variables were assumed to be sufficiently smooth in order to evaluate their derivatives. However, the presence of multiple phases leads to discontinuity of variables across the interface separating the phases. A complete description of the balance laws for multiphase materials is outside the scope of this book. However, we remark that the balance laws described in Sections 3.3.2-3.3.6 are valid within each individual phase. To describe the overall material behaviour, additional conditions for the interfaces have to be written. These conditions are referred to as interface *jump conditions*. Therefore, combinations of balance laws for each phase and jump conditions become the governing equations for material behaviour. Additionally, balances for each component can also be written. For example, we could write component mass balance (see Exercise 3.12).

The governing equations for multiphase and multicomponent materials will be more complex than the governing equations for single phase and single component materials. Therefore, simplifications are often made to reduce the complexity of the multiphase and multicomponent materials. Here, we mention a phenomenological simplification, which involves *homogenization* of the governing equations. Homogenization implies the assumption of an *effective single phase single component* behaviour from the complex behaviour. The balance laws and the constitutive relations used for complex materials are simplified with *effective parameters and variables*. These effective parameters and variables are estimated by a suitable combination of single phase parameters and variables. Even though the detailed treatment of multiphase and multicomponent materials is beyond the scope of this book, the homogenized treatment will be discussed in the book. It is always useful to work with a concept called *chemical potential* while dealing with multiphase and multicomponent materials, even when homogenized treatment is used. This concept will be developed below in Section 3.3.9.1.

3.3.9.1 Chemical Potential

We will first describe the definition of density and energy variables for a multicomponent material. The density of a component ρ_i is normally defined as the mass of the component i per unit volume of the material. Hence we can state that

$$\sum_i \rho_i = \rho . \tag{3.3.53}$$

To include the effect of composition on internal energy, Eqn. (3.3.39) is modified as follows:

$$\varepsilon = \varepsilon\left(\eta, \kappa_v, \rho_i\right). \tag{3.3.54}$$

Similar to Eqn. (3.3.40), we consider variation of the internal energy ε as

$$d\varepsilon = \left(\frac{\partial \varepsilon}{\partial \eta}\right)_{\kappa_v, \rho_i} d\eta + \left(\frac{\partial \varepsilon}{\partial \kappa_v}\right)_{\eta, \rho_i} d\kappa_v + \left(\frac{\partial \varepsilon}{\partial \rho_i}\right)_{\eta, \theta} d\rho_i . \tag{3.3.55}$$

Similar to Eqns. (3.3.41) and (3.3.42), we identify the additional partial derivative as the *chemical potential*

$$\left(\frac{\partial \varepsilon}{\partial \rho_i}\right)_{\eta, \theta} = \mu_i. \tag{3.3.56}$$

The chemical potential can be defined based on any of the energy measures, and therefore we have

$$\left(\frac{\partial \varepsilon}{\partial \rho_i}\right)_{\eta, \theta} = \left(\frac{\partial \psi}{\partial \rho_i}\right)_{\kappa_v, \theta} = \left(\frac{\partial \zeta}{\partial \rho_i}\right)_{\kappa_v, s_v} = \left(\frac{\partial \xi}{\partial \rho_i}\right)_{\eta, \kappa_v} = \mu_i . \tag{3.3.57}$$

Chemical potential can be visualized as the potential that generates the driving force for matter exchange. The use of chemical potential is similar to the use of other potentials that are familiar in mechanics and thermodynamics. For example, we know that a gradient of velocity potential causes fluid flow and a gradient in temperature (which is a thermal potential) causes heat flux. Therefore, the gradient of chemical potential will be useful in describing matter exchange processes such as diffusion, reaction and phase change.

3.4 CONSTITUTIVE RELATIONS

As we had seen in the previous section, the constitutive relations provide information about the material that will help us determine the field variables as solutions to the balance laws. In many approaches, the *constitutive relations* are implicitly stated and are substituted in the balance laws, in order to obtain *field equations* involving a particular problem. Many classical equations in fluid mechanics, such as the Navier Stokes equations, can be classified as field equations. This is a specialized equation that assumes the use of a Newtonian fluid, and is used in a wide range of boundary value problems. In a similar fashion, the biharmonic equation used in linearized elasticity assumes the use of a homogeneous, isotropic medium.

All the equations that are used in continuum problems must satisfy some properties of *invariance*. This means that they should describe a physical phenomenon in a unique way, yielding unique results, which are not dependent upon the way the observations are made. This uniqueness can be ensured if we deal with variables (vectors or tensors), which are described uniquely, and will not depend upon the way they are measured. Unfortunately, this cannot always be ensured immediately, especially when we are dealing with finite deformations and times. As an example, if we consider a vector like velocity, which is the rate of change of position with time, we know that the measured changes will depend upon the frame with respect to which the observations are made. Two observers from two different moving frames will observe the same phenomenon to have different velocities. Hence, the uniqueness of the measurement of an event called velocity can be ensured only if we are able to relate the different observed velocities, to each other, uniquely, through the relative movement of the reference frames. This statement about uniqueness of an observed quantity is called as *objectivity* of the observation and is always defined with respect to the transformations of the reference frames that are used to define this property. Quantities which satisfy objectivity are referred to as *objective* with respect to frame transformations. Objectivity of quantities is shown mathematically by showing their *frame invariance* or *frame indifference* with respect to transformations, as will be explained in Section 3.4.1. Additionally, objective quantities are referred to as *frame invariant* or *frame indifferent* quantities.

In the following discussion, we will examine the two popular ways of defining this property mathematically. We will also examine the implication of these properties on the structure of balance laws as well as the constitutive relations that could be defined for various materials.

3.4.1 Transformations

3.4.1.1 Euclidean Transformations

We had seen in Section 3.2.2.2 that rigid body motions of a body are characterized by the transformation that is given by (Eqn. (3.2.27))

$$\mathbf{x} = \mathbf{QX} + \mathbf{b}. \tag{3.4.1}$$

In the above transformation, let us assume that $\mathbf{X}$ refers to any distance vector between two points in the body. Further, let us assume that $\mathbf{Q}(t)$ refers to the rigid body rotations of an observer and $\mathbf{b}(t)$ refers to the rigid body translation of the moving reference frame. Then, denoting $\mathbf{x}'(t)$ as the observation of the vector $\mathbf{X}$ from the moving reference frame, we have

$$\mathbf{x}'(t) = \mathbf{Q}(t)\,\mathbf{X} + \mathbf{b}(t). \qquad (3.4.2(a))$$

While Eqn. 3.4.2(a) illustrates how the observation of a position vector may change due to rigid body motions, we could similarly foresee the observation of time changing due to rigid body motions. Hence, we could postulate that the observation of time from a moving reference frame (t') could be related to the observation of time in the fixed reference frame (t) through the relation :

$$t' = t + a, \qquad (3.4.2(b))$$

where a is an arbitrary fixed time interval. The transformations that are implied by Eqns. (3.4.2 (a), (b)), which essentially show how the measurement of distance and time can change due to motions of a reference frame, are called as *Euclidean transformations*.

3.4.1.2 Galilean Transformation

A particular type of Euclidean transformation, where the rigid body rotation (of observer) tensor $\mathbf{Q}$ is independent of time and rigid translation vector $\mathbf{b}(t)$ is through a time independent velocity vector $\mathbf{V}$, so that $\mathbf{b}(t) = \mathbf{V}t + \mathbf{c}$, will transform Eqn. (3.4.2) into

$$\mathbf{x}'(t) = \mathbf{Q}X + \mathbf{V}t + \mathbf{c} \ \text{ and } \ t' = t + a. \qquad (3.4.3)$$

Transformations of observer frame implied by Eqn. (3.4.3), which are special class of Euclidean transformations, are called as *Galilean transformations*. Both Euclidean and Galilean transformations are especially relevant in defining the objectivity of an observed event. It turns out that Galilean transformations are easier to satisfy for any observed quantity, compared to a more general Euclidean transformation. Objectivity with respect to either of these transformations are important while constructing any mathematical relations – either for the balance laws or for the constitutive relations. We will discuss this further after defining the statement of objectivity for various classes of vectors and tensors as discussed in the following section.

3.4.2 Objectivity of Mathematical Quantities

Let $\mathbf{e}_i$ represent the unit vectors in a chosen frame of reference and $\mathbf{f}_i$ represent the unit vectors in a rotated frame of reference. $\mathbf{f}_i$ are obtained by rotating $\mathbf{e}_i$ using an orthonal tensor $\mathbf{Q}(t)$, so that we have

$$\mathbf{f}_i(t) = \mathbf{Q}(t)\,\mathbf{e}_i. \qquad (3.4.4)$$

This is illustrated schematically for two dimensions in Fig. 3.14.

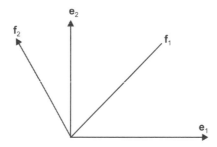

Fig. 3.14 Rotation of a reference frame

We will now denote all quantities with a superscript * to denote quantities that are measured with respect to the **f** co-ordinate frame and all quantities without a superscript to denote their measurement in the **e** co-ordinate frame. Let α be a scalar quantity as measured in the **e** co-ordinate frame. Since description of a scalar is independent of the co-ordinate frame, we have

$$\alpha^* = \alpha. \tag{3.4.5}$$

Consider a vector **a** measured in the **e** co-ordinate frame, which gets denoted as **a***, when measured in the **f** co-ordinate frame. The components a_i in the **e** co-ordinate frame will be given by

$$a_i = \mathbf{a} \cdot \mathbf{e}_i. \tag{3.4.6(a)}$$

Similarly, the components a^*_i of the vector **a*** along the f_i direction is given by

$$a^*_i = \mathbf{a}^* \cdot \mathbf{f}_i. \tag{3.4.6(b)}$$

Substituting for f_i from Eqn. (3.4.4) into (b) above, we get

$$a^*_i = \mathbf{a}^* \cdot \mathbf{Q}(t)\,\mathbf{e}_i = \mathbf{Q}^T(t)\mathbf{a}^* . \mathbf{e}_i \tag{3.4.6(c)}$$

From Eqn. (3.4.5), we have seen that the scalar components, as measured with respect to any co-ordinate frame, should remain the same. Hence, we must have

$$a_i = a^*_i \tag{3.4.6(d)}$$

Therefore, we get

$$\mathbf{a}^* = \mathbf{Q}(t)\,\mathbf{a}. \tag{3.4.7}$$

Equation (3.4.7) relates the measurements of vectors measured in the rotated co-ordinate frame **f** and the unrotated co-ordinate frame **e**. In other words, it establishes the conditions under which a measured vector will be objective, when the reference frames are subjected to *Euclidean transformations*.

If we have a tensor **T** as measured in the **e** co-ordinate frame which will be measured as **T*** in the **f** co-ordinate frame, it will be possible to prove (using component decomposition as illustrated for vector transformations above) that

$$\mathbf{T}^* = \mathbf{Q}\mathbf{T}\mathbf{Q}^{\mathrm{T}}. \tag{3.4.8}$$

Equation (3.4.8) expresses the conditions under which tensors measured with respect to two reference frames must be related to each other, if tensors are *frame indifferent* with respect to *Euclidean transformations* (see Exercise 3.17).

3.4.3 Invariance of Motions and Balance Equations

In the previous section, we examined the conditions of invariance for scalars, vectors and tensors, when they are subjected to Euclidean transformations. When we consider the most common vectors that are in use in mechanics we find that velocities and accelerations are not objective with respect to Euclidean transformations. However, it can be shown that the accelerations are invariant with respect to Galilean transformations, but velocities are not invariant with respect to Galilean transformations.

Using the properties of simple objects like vectors and tensors, it is also possible to characterize the nature of the partial differential equations where they appear. The invariant equations that will appear in all descriptions in continuum mechanics, are the balance equations.It has been shown that the balance equations are Galilean invariant. The constitutive equations, which are mathematical relations that are constructed in order to provide the necessary conditions to solve for the kinematic variables, are always used in conjunction with the balance equations. Hence, these relations must be at least Galilean invariant, so as to be compatible with the balance equations. In fact, most of the elementary material models that are used in fluid mechanics such as Newtonian material models, which are normally combined with balance equations to yield field equations, are Galilean invariant.

3.4.4 Invariance of Constitutive Relations

Since the constitutive relations are mathematical relations that are constructed by the user, they need to have a structure that will satisfy the invariance of some kind. A requirement of invariance of constructed constitutive relations is significant since we are dealing with motions that are large (in space and time), where the material can undergo large distortions whose description can be mathematically complex. Since these relations will be used in conjunction with the balance laws (which are Galilean invariant), the relations must at least be Galilean invariant. However, many constitutive relations that have been constructed for materials in the recent past have satisfied frame invariance with respect to Euclidean transformations. This means that if one envisages a particular relations between a deformation gradient $\mathbf{F}^*$ as measured in a moving frame of reference and $\mathbf{F}$ measured with respect to a fixed reference frame such that

$$\mathbf{F}^* = \mathbf{Q}(t)\mathbf{F}, \qquad (3.4.9)$$

where $\mathbf{Q}(t)$ is an orthogonal tensor. Then, all other relationships that are described in the constitutive relations must satisfy the objective criteria as described in Eqns. (3.4.5–3.4.7). Based on Eqn. (3.4.9), it can be shown (Example 3.4.1, Exercise 3.18) that velocity gradient tensor and spin tensors do not transform according to Eqn. (3.4.8), but the stretching tensor or strain rate tensor does, *viz.*

$$\left.\begin{array}{l} \mathbf{L}^* = \mathbf{Q}\mathbf{L}\mathbf{Q}^{\mathrm{T}} + \dot{\mathbf{Q}}\mathbf{Q}^{\mathrm{T}}, \\[2mm] \mathbf{L}^* = \mathbf{Q}\mathbf{W}\mathbf{Q}^{\mathrm{T}} + \dot{\mathbf{Q}}\mathbf{Q}^{\mathrm{T}}, \\[2mm] \mathbf{D}^* = \mathbf{Q}\mathbf{D}\mathbf{Q}^{\mathrm{T}}. \end{array}\right\} \qquad (3.4.10)$$

Example 3.4.1: Show that $\mathbf{L}^*$, the velocity gradient measured in the **f** coordinate frame, is related to $\mathbf{L}$ as follows:

$$\mathbf{L}^* = \mathbf{Q}\mathbf{L}\mathbf{Q}^{\mathrm{T}} + \dot{\mathbf{Q}}\mathbf{Q}^{\mathrm{T}}$$

Solution:

We know from the definition of velocity gradient that (Example 3.2.3)

$$\mathbf{L} = \dot{\mathbf{F}}\mathbf{F}^{-1}$$

Using Eqn. (3.4.8) we can find $\dot{\mathbf{F}}^{-1^*}$ and $\mathbf{F}^{-1^*}$, measures in the **f** coordinate frame

$$\dot{\mathbf{F}}^* = \dot{\mathbf{Q}}\mathbf{F} + \mathbf{Q}\dot{\mathbf{F}}$$

Similarly, we can show that

$$\mathbf{F}^{-1^*} = \mathbf{F}^{-1}\mathbf{Q}^T$$

Therefore, $\mathbf{L}^*$ can be written as

$$\mathbf{L}^* = \dot{\mathbf{F}}^*\mathbf{F}^{-1^*} = \left(\dot{\mathbf{Q}}\mathbf{F} + \mathbf{Q}\dot{\mathbf{F}}\right)\mathbf{F}^{-1}\mathbf{Q}^T$$

$$= \dot{\mathbf{Q}}\mathbf{F}\mathbf{F}^{-1}\mathbf{Q}^T + \mathbf{Q}\dot{\mathbf{F}}\mathbf{F}^{-1}\mathbf{Q}^T$$

$$= \dot{\mathbf{Q}}\mathbf{Q}^T + \mathbf{Q}\mathbf{L}\mathbf{Q}^T$$

The above equation tells us that $\mathbf{L}$ is not objective with respect to orthogonal transformations. Hence, it is not advisable to use it in describing constitutive relations.

Based on Eqns. (3.4.8) and (3.4.10), we can state that $\mathbf{D}$ is objective, while $\mathbf{L}$ and $\mathbf{W}$ are not. We will observe later that $\mathbf{D}$ will be included in the constitutive relations of various materials. We will illustrate the concepts more clearly through discussion on two specific class of materials, thermoelastic and thermoviscous, in following sections.

3.4.4.1 Frame Invariance in Thermoelastic Material

We will consider a thermoelastic material where the stresses σ, internal energy e and heat flux $\mathbf{q}$ are some functions of the deformation gradient $\mathbf{F}$, temperature θ and gradient of temperature, grad θ. Hence, we have

$$\sigma = \hat{\sigma}(\mathbf{F}, \theta, \operatorname{grad}\theta)$$

$$\varepsilon = \hat{\varepsilon}(\mathbf{F}, \theta, \operatorname{grad}\theta) \text{ and}$$

$$\mathbf{q} = \hat{\mathbf{q}}(\mathbf{F}, \theta, \operatorname{grad}\theta) \tag{3.4.10(a)}$$

In addition, we will also use the second law of thermodynamics, which was stated as an inequality (Eqn. (3.3.36)). In doing so, we have

$$\eta = \hat{\eta}(\mathbf{F}, \theta, \operatorname{grad}\theta) \tag{3.4.10(b)}$$

We will now examine as to how these relations need to be transformed if the relations are to remain objective with respect to Euclidean transformations. This means that we have to examine the effect of a transformation given in Eqn. (3.4.9). As an example, we will consider a particular form of the internal energy equation given by

$$\hat{\varepsilon}(\mathbf{F}) = \operatorname{tr}(\mathbf{F}^T\mathbf{F}) \tag{3.4.11}$$

If we substitutive the transformation of reference frames implied by Eqn. (3.4.10) into Eqn. (3.4.11), we get

$$\hat{\varepsilon}(\mathbf{F}^*) = \operatorname{tr}(\mathbf{F}^{*^T}\mathbf{F}^*) = \operatorname{tr}(\mathbf{F}^T\mathbf{Q}^T\mathbf{Q}\mathbf{F}) = \operatorname{tr}(\mathbf{F}^T\mathbf{F}) = \hat{\varepsilon}(\mathbf{F})$$
$$\tag{3.4.12}$$

Equation (3.4.12) illustrates that the expression for internal energy as given in Eqn. (3.4.11) remains invariant with regard to orthogonal transformation of the co-ordinate frames as indicated by Eqn. (3.4.4). Since the internal energy is a scalar, from the criteria established in Eqn. (3.4.5), we can say that constitutive relation suggested by Eqn. (3.4.11) is objective, with respect to Euclidean transformations. It is easy to prove (by doing a similar exercise) that a constitutive relation of the type $\hat{\varepsilon}(\mathbf{F}) = \operatorname{tr}(\mathbf{F})$ *will not* be objective with respect to Euclidean transformations.

We can pick the orthogonal transformations to be of the type

$$\mathbf{Q} = \mathbf{R}^T. \tag{3.4.13}$$

If we substitute Eqn. (3.4.13) in Eqn. (3.4.12), we get

$$\hat{\varepsilon}(\mathbf{F}^*) = \hat{\varepsilon}(\mathbf{QF}) = \hat{\varepsilon}(\mathbf{R}^T\mathbf{RU}) = \hat{\varepsilon}(\mathbf{U}) = \hat{\varepsilon}(\mathbf{F}) \qquad (3.4.14)$$

Equation (3.4.14) tells us that as a consequence of having a constitutive relation that is objective with respect to Euclidean transformations, it is possible to express the internal energy of an elastic material to be purely in terms of the right stretch tensor $\mathbf{U}$.

If we were to examine the conditions for objectivity under Euclidean transformations of the constitutive relations for the Cauchy stresses σ, we would find that they should satisfy the conditions implied by Eqn. (3.4.7), when the relations are subjected to the transformations implied by Eqn. (3.4.8). This would mean that the expressions for σ should satisfy

$$\hat{\sigma}(\mathbf{F}^*) = \hat{\sigma}(\mathbf{QF}) = \mathbf{Q}\hat{\sigma}(\mathbf{F})\mathbf{Q}^T . \qquad (3.4.15)$$

Which can also be re-written as

$$\hat{\sigma}(\mathbf{F}) = \mathbf{Q}^T\hat{\sigma}(\mathbf{QF})\mathbf{Q} . \qquad (3.4.16)$$

Equation (3.4.16) tells us the necessary conditions that must be satisfied by the constitutive equation for stresses in an elastic material, if the relation is to be objective with respect to Euclidean transformation. As before, again we can choose $\mathbf{Q} = \mathbf{R}^T$, so that we have, as a consequence of objectivity, the following relation:

$$\hat{\sigma}(\mathbf{F}) = \mathbf{R}\hat{\sigma}(\mathbf{U})\mathbf{R}^T . \qquad (3.4.17)$$

The restrictions imposed by Eqn. (3.4.17) gives us the constraints on the form that the constitutive equation can take. Sometimes, the restrictions imposed by Eqn. (3.4.17), or similar equations for any general material will become too restrictive, thereby complicating the form of the constitutive equation used for the material. In many such cases, it is recommended that the constitutive equation needs only to satisfy the Galilean invariance, which is satisfied by the balance laws and not the Euclidean frame invariance, which is discussed above. A more detailed discussion on these topics is beyond the scope of the current text. In the current treatment, we will only appreciate the importance of this requirement and start examining many popular and simplistic models in the literature, which are known to satisfy the material frame invariance.

3.4.4.2 Constitutive Relations for Thermoelastic Materials

We have seen in the previous section that the constitutive relations for a thermoelastic material can be reduced to the form

$$\sigma = \mathbf{R}\hat{\sigma}(\mathbf{U}, \theta, \text{grad } \theta)\mathbf{R}^T ,$$

$$\varepsilon = \hat{\varepsilon}(\mathbf{U}, \theta, \text{grad } \theta)$$

$$\mathbf{q} = \hat{\mathbf{q}}(\mathbf{U}, \theta, \text{grad}\,\theta) \text{ and}$$

$$\eta = \hat{\eta}(\mathbf{U}, \theta, \text{grad}\,\theta). \tag{3.4.18}$$

It is also possible to express the constitutive relations for thermoelastic material in the reference configuration, so that we have

$$\mathbf{S} = \hat{\mathbf{S}}(\mathbf{F}, \theta, \text{Grad}\,\theta)$$

$$\psi = \hat{\psi}(\mathbf{F}, \theta, \text{Grad}\,\theta)$$

$$\eta = \hat{\eta}(\mathbf{F}, \theta, \text{Grad}\,\theta) \text{ and}$$

$$\mathbf{q} = \hat{\mathbf{q}}(\mathbf{F}, \theta, \text{Grad}\,\theta) \tag{3.4.19}$$

In writing the referential description in Eqn. (3.4.19), we have chosen to express the Helmhotlz free energy ψ defined by $\psi = \varepsilon - \eta\theta$ (Eqn. (3.3.51)), instead of the internal energy ε. As discussed in Section 3.2.6.3, the choice of state variable is arbitrary. However, it is found in practice that the choice of ψ helps in simplifying the associated derivations since ψ can be directly related to the strain energy density that is defined for elastic materials.

Constraints Imposed by Entropy Inequality

Let us examine closely the entropy inequality of Eqn. (3.3.37). This equation is reproduced below in a slightly modified form:

$$\dot{\eta} - \frac{\text{div}\left(\dfrac{\mathbf{q}}{\theta}\right)}{\rho} - \frac{r}{\theta} \geq 0 \tag{3.4.20}$$

or $\qquad \gamma \geq 0; \text{ where } \gamma = \dot{\eta} - \dfrac{\text{div}\left(\dfrac{\mathbf{q}}{\theta}\right)}{\rho} - \dfrac{r}{\theta} \qquad (3.4.21)$

Substituting for heat generation rate γ from Eqn. (3.3.35) into Eqn. (3.4.21), it is possible to show that

$$\gamma = \frac{\mathbf{q} \cdot \text{grad}\,\theta}{\rho\theta^2} + \frac{\sigma : \mathbf{D}}{\rho\theta} - \frac{\dot{\psi}}{\theta} - \frac{\eta\dot{\theta}}{\theta} \tag{3.4.22}$$

We now substitutive for internal energy ε in terms of Helmoltz free energy ψ so as to obtain an expression for γ in the deformed configuration, or in the reference configuration, as shown below:

$$\left. \begin{aligned} \gamma &= \frac{\mathbf{q} \cdot \text{grad}\,\theta}{\rho\theta^2} + \frac{\sigma : \mathbf{D}}{\rho\theta} - \frac{\dot{\psi}}{\theta} - \frac{\eta\dot{\theta}}{\theta} \\[2mm] &= \frac{\mathbf{q}_o \cdot \text{Grad}\,\theta}{\rho_o\theta^2} + \frac{\mathbf{S} : \dot{\mathbf{F}}}{\rho_o\theta} - \frac{\dot{\psi}}{\theta} - \frac{\eta\dot{\theta}}{\theta} \end{aligned} \right\} \tag{3.4.23}$$

We will now substitutive a notation $\mathbf{g} = \text{Grad } \theta$, expanding ψ in terms of its independent variables using chain rule, and rearrange the terms in Eqn. (3.4.23) to obtain the following expression:

$$\gamma = \frac{1}{\rho_o \theta}\left(\hat{\mathbf{S}} - \rho_o \frac{\partial \hat{\psi}}{\partial \mathbf{F}}\right):\dot{\mathbf{F}} - \left(\frac{\partial \hat{\psi}}{\partial \theta} + \hat{\eta}\right)\frac{\dot{\theta}}{\theta} - \frac{\partial \hat{\psi}}{\partial \mathbf{g}}\cdot\frac{\dot{\mathbf{g}}}{\theta} + \frac{\mathbf{q}_o \cdot \mathbf{g}}{\rho_o \theta^2} \geq 0. \qquad (3.4.24)$$

We will now examine the conditions under which the inequality Eqn. (3.4.24) will be unconditionally satisfied for all possible variation of the independent variables. In other words, the terms with $\dot{\mathbf{F}}$, θ and $\dot{\mathbf{g}}$ should satisfy the inequality for any arbitrary and independent variation in $\dot{\mathbf{F}}$, θ and $\dot{\mathbf{g}}$, respectively. One of the ways of ensuring that this will happen is to insist that these terms will always be zero. We should note that this insistence is possible because we have assumed that (in Eqn. (3.4.19)) the stress and Helmholtz free energy are not to be fuctions of $\dot{\mathbf{F}}$, θ and $\dot{\mathbf{g}}$. Hence, we can derive the following relations between the various variables in a thermoelastic material:

$$\mathbf{S} = \rho_o \frac{\partial \hat{\psi}}{\partial \mathbf{F}},$$

$$\eta = -\frac{\partial \hat{\psi}}{\partial \theta},$$

$$\frac{\partial \hat{\psi}}{\partial \mathbf{g}} = \mathbf{0},$$

$$\frac{\mathbf{q}_o \cdot \mathbf{g}}{\rho_o \theta^2} \geq 0 \qquad (3.4.25)$$

We also note that no deduction can be made about the term containing heat flux and temperature gradient ($\mathbf{q}$ is a function of $\mathbf{g}$). Therefore, we maintain that this term should satisfy the inequality. Equation (3.4.25) also tell us that the Helmholtz free energy will not be a function of the gradient of temperature $\mathbf{g}$. Hence, this can be eliminated as a variable in $\emptyset$ and therefore $\mathbf{S}$ (since $\mathbf{S}$ is a function of $\emptyset$, Eqn. (3.4.25)). Therefore, the constitutive relations for a themoelastic material will be

$$\psi = \hat{\psi}(\mathbf{F}, \theta),$$

$$\mathbf{S} = \rho_o \frac{\partial \hat{\psi}(\mathbf{F}, \theta)}{\partial \mathbf{F}}, \qquad (3.4.26)$$

$$\eta = -\frac{\partial \hat{\psi}(\mathbf{F}, \theta)}{\partial \theta}$$

Equation for heat flux remains unaltered (Eqn. (3.4.19)) since $\mathbf{q}$ is not a function of $\emptyset$. We note that a general elastic material will be a special case of a thermoelastic material, where the properties are defined at a particular temperature. Hence, Eqn. (3.4.26) tells us that it is possible to obtain the stress field in a general non-linear elastic material, provided we are able to define a Helmholtz free energy function for the material. Further, from Eqn. (3.4.14), it is easy for us to see that objectivity of internal energy will help us to replace $\mathbf{F}$ with $\mathbf{C}$ in all the relations of Eqn. (3.4.26), so that the constitutive relations can be written as

$$\psi = \hat{\psi}(\mathbf{C}, \theta),$$

$$\mathbf{S} = 2\rho_o \mathbf{F} \frac{\partial \hat{\psi}(\mathbf{C}, \theta)}{\partial \mathbf{C}},$$

$$\eta = -\frac{\partial \hat{\psi}(\mathbf{C}, \theta)}{\partial \theta} \qquad (3.4.27)$$

In nonlinear elasticity, a special form of the Helmholtz free energy function, called as the *strain energy density function W* is defined, and this is related to Helmholtz free energy through the relation

$$W = \rho_o \psi. \qquad (3.4.28)$$

We note that the strain energy density function can be defined as the Helmholtz free energy per unit volume. In nonlinear elasticity, we normally work only with a possible expressions for W and derive the kinetic quantities from an assumed form of W. More of these details will be dealt with in Chapter 6 on models on nonlinear elastic solids later.

3.4.4.3 Frame Invariance and Constitutive Relations for a Thermoviscous Fluid

In Sections 3.4.4.1 and 3.4.4.2, we discussed the frame invariance and constitutive models for a thermoelastic material. In this section, we will describe similar considerations for thermoviscous fluids. Since the discussion is similar to that in Sections 3.4.4.1 and 3.4.4.2, we highlight only the main steps.

Consider a thermoviscous material (Newtonian fluid is a common example of this class of materials), for which stress σ, internal energy ε, entropy η and heat flux $\mathbf{q}$ are some functions of deformation gradient $\mathbf{F}$, temperature θ and gradient of temperature, grad θ. Additionally, we also assume the dependence on the rate of deformation gradient $\dot{\mathbf{F}}$ and specific volume v. Similar to the discussion of Eqns. (3.4.11)-(3.4.17), we can show that frame invariance requires that both internal energy and stresses be functions of $\mathbf{U}$ and $\dot{\mathbf{U}}$, in place of $\mathbf{F}$ and $\dot{\mathbf{F}}$, respectively. Therefore, we can write

$$\sigma = \mathbf{R}\hat{\sigma}(\mathbf{U},\dot{\mathbf{U}}, \ v, \ \theta, \text{grad } \theta)\mathbf{R}^T ,$$

$$\varepsilon = \hat{\varepsilon}(\mathbf{U},\dot{\mathbf{U}}, \ v, \ \theta, \text{grad } \theta) ,$$

$$\mathbf{q} = \hat{\mathbf{q}}(\mathbf{U},\dot{\mathbf{U}}, \ v, \ \theta, \text{grad } \theta) ,$$

$$\eta = \hat{\eta}(\mathbf{U},\dot{\mathbf{U}}, \ v, \ \theta, \text{grad } \theta) . \qquad (3.4.29)$$

As was mentioned in Section 3.2.6, relative kinematical measures are used for fluid-like materials. Therefore, Eqn. (3.4.29) is rewritten in terms of $\mathbf{U}_t$ and $\dot{\mathbf{U}}_t$. We recall that these quantities are evaluated at the present time and we have $\mathbf{U}_t = \mathbf{I}$ and $\dot{\mathbf{U}}_t = \mathbf{D}$. Therefore, Eqn. (3.4.29) can be written as

or
$$\sigma = \mathbf{R}\hat{\sigma}(\mathbf{D}, \ v, \ \theta, \text{grad } \theta)\mathbf{R}^T ,$$

$$\varepsilon = \hat{\varepsilon}(\mathbf{D}, \ v, \ \theta, \text{grad } \theta) ,$$

or
$$\psi = \hat{\psi}(\mathbf{D}, \ v, \ \theta, \text{grad } \theta) ,$$

$$\mathbf{q} = \hat{\mathbf{q}}(\mathbf{D}, \ v, \ \theta, \text{grad } \theta) \text{ and}$$

$$\eta = \hat{\eta}(\mathbf{D}, \ v, \ \theta, \text{grad } \theta) . \qquad (3.4.30)$$

Constraints Imposed by Entropy Inequality

Starting from Eqn. (3.4.23) written in terms of current configuration and substituting for $\hat{\psi}$ from Eqn. (3.4.30), we obtain (similar to Eqn. (3.4.24))

$$\gamma = \frac{1}{\rho\theta}(\sigma:\mathbf{D}) - \frac{1}{\theta}\left(\frac{\partial\hat{\psi}}{\partial\mathbf{D}}:\dot{\mathbf{D}}\right) - \frac{1}{\rho\theta}\frac{\partial\hat{\psi}}{\partial v}(\text{div } \mathbf{v}) -$$

$$\left(\frac{\partial\hat{\psi}}{\partial\theta}+\hat{\eta}\right)\frac{\dot{\theta}}{\theta} - \frac{\partial\hat{\psi}}{\partial\mathbf{g}}\cdot\frac{\dot{\mathbf{g}}}{\theta}+\frac{\mathbf{q}\cdot\mathbf{g}}{\rho\theta^2} \geq 0 \quad (3.4.31)$$

where we have used the balance of mass (Eqn. (3.3.7)) to evaluate $\dot{v}$ and $\dot{\rho}$. Rearranging Eqn. (3.4.30), we get,

$$\gamma = \frac{1}{\rho\theta}\left(\sigma - \frac{\partial\hat{\psi}}{\partial v}\mathbf{I}\right):\mathbf{D} - \frac{1}{\theta}\left(\frac{\partial\hat{\psi}}{\partial\mathbf{D}}:\dot{\mathbf{D}}\right) - \frac{1}{\theta}\left(\frac{\partial\hat{\psi}}{\partial\mathbf{g}}\cdot\dot{\mathbf{g}}\right) -$$

$$\frac{1}{\theta}\left(\frac{\partial\hat{\psi}}{\partial\theta}+\hat{\eta}\right)\dot{\theta}+\frac{\mathbf{q}\cdot\mathbf{g}}{\rho\theta^2} \geq 0 . \quad (3.4.32)$$

We can use arguments similar to those used to obtain deductions from Eqn. (3.4.24) to simplify Eqn. (3.4.32). Since Eqn. (3.4.32) must hold for all variations of independent variables $\dot{\mathbf{D}}$, $\dot{\theta}$ and $\dot{\mathbf{g}}$, we can state that

$$\eta = -\frac{\partial \hat{\psi}}{\partial \theta},$$

$$\frac{\partial \hat{\psi}}{\partial \mathbf{g}} = 0,$$

$$\frac{\partial \hat{\psi}}{\partial \mathbf{D}} = 0 \qquad\qquad (3.4.33)$$

and $\quad \frac{1}{\rho\theta}\left(\boldsymbol{\sigma} - \frac{\partial \hat{\psi}}{\partial v}\mathbf{I}\right):\mathbf{D} + \frac{\mathbf{q}\cdot\mathbf{g}}{\rho\theta^2} \geq 0.$

We cannot make any deductions about the first and the last terms of Eqn. (3.4.32), because $\boldsymbol{\sigma}$ is a function of $\mathbf{D}$ and $\mathbf{q}$ is a function of $\mathbf{g}$. Equation (3.4.33) tells us that Helmholtz free energy is not a function of $\mathbf{g}$ and $\mathbf{D}$. Therefore, we can write

$$\psi = \hat{\psi}(v, \theta). \qquad\qquad (3.4.34)$$

Based on Table 3.2 and similar to Eqns. (3.3.41) and (3.3.42), we define the *pressure* as

$$p = -\frac{\partial \hat{\psi}}{\partial v} \qquad\qquad (3.4.35)$$

where, the negative sign is due to the convention that compressive pressure is positive.

Therefore, we can simplify the inequality in Eqn. (3.4.32) as

$$\frac{1}{\rho\theta}(\boldsymbol{\sigma} + p\mathbf{I}):\mathbf{D} + \frac{\mathbf{q}\cdot\mathbf{g}}{\rho\theta^2} \geq 0 \qquad\qquad (3.4.36)$$

We define the *deviatoric stress* ($\boldsymbol{\tau}$) measure as

$$\boldsymbol{\tau} = (\boldsymbol{\sigma} + p\mathbf{I})$$

or $\qquad\qquad\qquad \boldsymbol{\sigma} = -p\mathbf{I} + \boldsymbol{\tau} \qquad\qquad (3.4.37)$

Equation (3.4.37) helps us to decompose the total stress into a volumetric stress (p) and deviatoric stress ($\boldsymbol{\tau}$). This decomposition is useful while describing constitutive relations of different materials, as will be shown in Chapters 4 and 5.

Finally, we can write the simplified form of Eqn. (3.4.30) based on the restrictions imposed by the second law as

$$\boldsymbol{\tau} = \mathbf{R}\hat{\boldsymbol{\tau}}(\mathbf{D}, v, \theta, \mathrm{grad}\,\theta)\mathbf{R}^T$$

$$p = (\theta, v),$$

$$\varepsilon = \hat{\varepsilon}(v,\ \theta) \text{ or } \psi = \hat{\psi}(v,\ \theta)$$

$$\eta = \hat{\eta}(\theta,\ v)$$

$$q = \hat{q}(\mathbf{D},\ v,\ \theta \text{ grad } \theta) \qquad (3.4.38)$$

It may be noted in Eqn. (3.4.38) that the internal energies ε and ψ, entropy η and pressure p, are independent of $\mathbf{D}$ and grad θ, while a similar generalization cannot be stated for the stresses τ or the heat flux $\mathbf{q}$. The independence of ε, ψ, η and p directly follow from Eqns. (3.4.33) and (3.4.35). Similar generalizations are not possible for τ and $\mathbf{q}$.

3.4.5 Frame Invariance of Derivatives

In Section 3.2.7, we described the Jaumann and convected derivatives. It was mentioned there that these derivatives are important in evaluation of rates of kinematic and kinetic variables. These derivatives are very useful because they are objective. In the following discussion, we show that the material derivative of a tensor is not objective with respect to frame transformations. However, the Jaumann derivative of a tensor will be shown to be frame invariant with respect to orthogonal transformations.

Based on Eqn. (3.4.7), a tensor $\mathbf{T}$ will be frame invariant if it transforms according to $\mathbf{T}^* = \mathbf{QTQ}^T$. Let us examine how material derivative $\dot{\mathbf{T}}$ transforms with respect to orthogonal transformations. We can define $\dot{\mathbf{T}}^*_{\frac{1}{2}}$ as the material derivative of $\mathbf{T}$ in f co-ordinate frame. Therefore, we obtain

$$\dot{\mathbf{T}}^* = \overparen{\dot{\mathbf{Q}\mathbf{T}\mathbf{Q}^T}} = \dot{\mathbf{Q}}\mathbf{T}\mathbf{Q}^T + \mathbf{Q}\dot{\mathbf{T}}\mathbf{Q}^T + \mathbf{Q}\mathbf{T}\dot{\overparen{\mathbf{Q}}}^T \qquad (3.4.39)$$

By rearranging Eqn. (3.4.39), we get

$$\dot{\mathbf{T}}^* = \mathbf{Q}\dot{\mathbf{T}}\mathbf{Q}^T + \left[\dot{\mathbf{Q}}\mathbf{T}\mathbf{Q}^T + \mathbf{Q}\mathbf{T}\dot{\overparen{\mathbf{Q}}}^T \right] \qquad (3.4.40)$$

Presence of additional terms in Eqn. (3.4.40) when comparing with Eqn. (3.4.7) shows that the material derivative $\dot{\mathbf{T}}$ is not objective with respect to orthogonal transformations. On the other hand, Jaumann derivative can be shown to be objective with respect to orthogonal transformations (see Example 3.4.2)

$$\mathbf{Q}\overset{\square}{\mathbf{T}}\mathbf{Q}^T = \overset{\square}{\mathbf{T}}{}^* \qquad (3.4.41)$$

where $\overset{\square}{\mathbf{T}}$ is the Jaumann derivative of $\mathbf{T}$ in the f co-ordinate. Similarly, it can be shown that the convected derivatives defined in Section 3.2.7 are also frame invariant (see Exercise 3.19).

Example 3.4.2: Show that Jaumann derivative is objective.

Solution:

As defined in Eqn. (3.2.73), the Jaumann derivative of a tensor T is

$$\overset{\square}{\mathbf{T}} = \dot{\mathbf{T}} + \mathbf{TW} + \mathbf{W}^T\mathbf{T}$$

Let us examine the orthogonal transformation of this derivative

$$\mathbf{Q}\overset{\square}{\mathbf{T}}\mathbf{Q}^T = \mathbf{Q}\left(\dot{\mathbf{T}} + \mathbf{TW} + \mathbf{W}^T\mathbf{T}\right)\mathbf{Q}^T$$

Substituting for $\mathbf{Q}\mathbf{T}\mathbf{Q}^T$ from Eqn. (3.4.40) in Eqn. (3.4.41), we get

$$\mathbf{Q}\overset{\square}{\mathbf{T}}\mathbf{Q}^T = \dot{\mathbf{T}}^* - \left[\dot{\mathbf{Q}}\mathbf{T}\mathbf{Q}^T + \mathbf{Q}\mathbf{T}\overset{\displaystyle\frown}{\dot{\mathbf{Q}}^T}\right] + \mathbf{Q}\left(\mathbf{TW} + \mathbf{W}^T\mathbf{T}\right)\mathbf{Q}^T$$

Our attempt will be to show that the right-hand side of the above equation is

$\overset{\square}{\mathbf{T}}^*$. We use the property of $\mathbf{Q}$ that $\mathbf{Q}^T\mathbf{Q} = \mathbf{I} = \mathbf{Q}\mathbf{Q}^T$ and apply it in the last four terms of the above equation:

$$\mathbf{Q}\overset{\square}{\mathbf{T}}\mathbf{Q}^T = \dot{\mathbf{T}}^* - \left[\dot{\mathbf{Q}}\mathbf{Q}^T\mathbf{Q}\mathbf{T}\mathbf{Q}^T + \mathbf{Q}\mathbf{T}\mathbf{Q}^T\overset{\displaystyle\frown}{\mathbf{Q}\dot{\mathbf{Q}}^T}\right]$$
$$+ \mathbf{Q}\left(\mathbf{T}\mathbf{Q}^T\mathbf{Q}\mathbf{W} + \mathbf{W}^T\mathbf{Q}^T\mathbf{Q}\mathbf{T}\right)\mathbf{Q}^T.$$

We recognize that $\mathbf{T}^* = \mathbf{Q}\mathbf{T}\mathbf{Q}^T$ (Eqn. (3.4.7)). Therefore, we get

$$\mathbf{Q}\overset{\square}{\mathbf{T}}\mathbf{Q}^T = \dot{\mathbf{T}}^* - \left[\dot{\mathbf{Q}}\mathbf{Q}^T\mathbf{T}^* + \mathbf{T}^*\overset{\displaystyle\frown}{\mathbf{Q}\dot{\mathbf{Q}}^T}\right] + \left(\mathbf{T}^*\mathbf{Q}\mathbf{W}\mathbf{Q}^T + \mathbf{Q}\mathbf{W}^T\mathbf{Q}^T\mathbf{T}^*\right)$$

Collecting the third and fourth terms (of the right-hand side from the above equation), we get

$$\mathbf{Q}\overset{\square}{\mathbf{T}}\mathbf{Q}^T = \dot{\mathbf{T}}^* + \mathbf{T}^*\left(\mathbf{Q}\mathbf{W}\mathbf{Q}^T + \dot{\mathbf{Q}}\mathbf{Q}^T\right) + \left(\mathbf{Q}\mathbf{W}^T\mathbf{Q}^T - \dot{\mathbf{Q}}\mathbf{Q}^T\right)\mathbf{T}^*$$

where we used the property of $\mathbf{Q}$ such that $\dot{\mathbf{Q}}\mathbf{Q}^T + \mathbf{Q}\overset{\displaystyle\frown}{\dot{\mathbf{Q}}^T} = 0$. Using the expression for $\mathbf{W}^*$ from Eqn. (3.4.9), we get

$$\mathbf{Q}\overset{\square}{\mathbf{T}}\mathbf{Q}^T = \dot{\mathbf{T}}^* + \mathbf{T}^*\mathbf{W}^* + \mathbf{W}^{T^*}\mathbf{T}^* = \overset{\square}{\mathbf{T}}^*$$

Therefore, we have shown that the Jaumann derivative of a tensor $\mathbf{T}$ is objective with respect to orthogonal transformations.

SUMMARY

The current chapter outlines the scope and contents of continuum mechanics. The kinematic measures for large deformations are introduced and their difference with measures normally used in infinitismal theory is outlined. The differences in measuring these quantities with respect to undeformed and deformed configurations are described. The balance laws that are mainly the application of Newton's laws and conservation laws for a general continuum, are derived. The thermodynamic balance laws as well as the entropy inequality, as applied to a continuum are derived. The need for the existence of constitutive equations for the completeness of the problem is established. The requirements of objectivity with respect to Euclidean and Galilean transformations, for these constitutive equations are also illustrated.

In this chapter, we also derived the possible forms of constitutive relations for thermoelastic and thermoviscous materials. We have seen that the statement of objectivity will help us in identifying the nature of the dependent variables in the constitutive relations. We have also seen that the thermodynamic restriction posed by the second law will help us to identify the nature of relationships that could exist in any chosen class of material. A detailed application of these derived forms for engineering materials will be illustrated in later chapters.

While the constitutive equations mentioned in this chapter could be applicable to any continuum which is bounded by a boundary, we will restrict our attention mainly to solids and liquids that are most commonly used in engineering. The broad classes of these constitutive equations, their general mathematical form under each class etc. will be discussed in more detail when we discuss both the broad types of materials, solid-like and fluid-like, in the subsequent chapters.

EXERCISE

3.1. Consider rigid rotation of a cube (side $= a$) around third axis (perpendicular to the plane of the paper) as shown in figure.

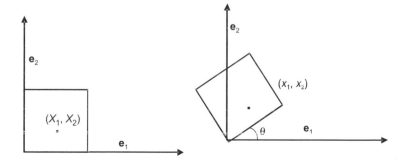

(a) Write x_i as functions of X_j, $x_i = x_i(X_j)$.

(b) Evaluate $\mathbf{F}$, eigenvalues of $\mathbf{F}$ and J.

(c) Argue that $\mathbf{U} = \mathbf{V} = \mathbf{I}$.

(d) Show that, as expected, $\mathbf{E} = \dfrac{1}{2}(\mathbf{FF}^T - \mathbf{I}) = \mathbf{0}$.

(e) Evaluate Grad $\mathbf{u}$.

(f) Show that $\mathbf{e}$ is non-zero, even though the above change in configuration is due to rigid-body rotation.

(g) Assume θ to be very small. Show that in this case, even for a rigid body motion, $\mathbf{e}$ will reduce to $\mathbf{0}$.

3.2. Show that

(a) $\operatorname{Div} \varphi = J \operatorname{div}\left(J^{-1}\mathbf{F}\cdot\varphi\right)$

(b) $\dfrac{d}{dt}\displaystyle\int_{D'} \phi\, dV_x = \int_{D'}\left(\dot\phi + \phi\operatorname{div}\mathbf{v}\right)dV_x$

3.3. Show that $\mathbf{Wm}\cdot\mathbf{m} = 0$ (Eqn. (3.2.47)) .

3.4. Show that

$$\mathbf{W} = \dot{\mathbf{R}}\mathbf{R}^T + \dfrac{1}{2}\mathbf{R}\left(\dot{\mathbf{U}}\mathbf{U}^{-1} - \mathbf{U}^{-1}\dot{\mathbf{U}}\right)\mathbf{R}^T$$

3.5. If $x_i = A_{ij}X_j + b_i$ and $\mathbf{A}$ is orthogonal tensor, evaluate $\mathbf{E}$.

3.6. For the following velocity fields, evaluate $\mathbf{L}$, $\mathbf{D}$ and $\mathbf{W}$:

(a) $\mathbf{v} = v_x\mathbf{e}_x$

(b) $\mathbf{v} = (k/r)\mathbf{e}_\theta$

(c) $\mathbf{v} = kr\mathbf{e}_\theta$

Is the path of a material particle related to $\mathbf{W}$ being zero or non-zero?

3.7. Show that D_{12} represents the rate of change of the angle between any two fibers that were originally oriented along the principal directions $\mathbf{e}_1$ and $\mathbf{e}_2$.

3.8. The flow field of flow of water out of a kitchen sink is given by the expression below :

$y_1 = (x_1 \cos\theta\,(r) - x_2 \sin\theta\,(r))\exp(-\alpha t/2)$

$y_2 = (x_1 \sin\,(r) + x_2 \cos\theta\,(r))\exp(-\alpha t/2)$

$y_3 = x_3 \exp(\alpha t)$

where $r = \sqrt{x_1^2 + x_2^2}$ is the distance from the x_3 axis and θ is some function of r.

Calculate the following:

(a) Components of the velocity **v** of a particle at the instant t.

(b) Div **v** and div v .

(c) Determine the particle paths and streamlines of this flow. Describe them qualitatively. Is this isochoric?

3.9. Show that acceleration of a material particle can be expressed as

$$\mathbf{a} = \frac{\partial \mathbf{v}}{\partial t} + \text{curl } \mathbf{v} \times \mathbf{v} + \frac{1}{2}\text{grad}\left(v^2\right)$$

3.10. (a) Current and reference configurations can be related to each other as follows (based on Eqns. (3.2.27) and (3.2.51)) :

$$\mathbf{x} = \mathbf{Q} \cdot \mathbf{X} + \mathbf{b}$$

$$x = \mathbf{Q}^T \mathbf{x} - \mathbf{Q}^T \mathbf{b}$$

Using these equations, show that velocity can be expressed in terms of a generalized body spin tensor $\mathbf{\Omega}(t)$ and a space independent tensor $\mathbf{C}(t)$ as (Eqn. (3.2.52))

(b) Since $\mathbf{\Omega}(t)$ is skew-symmetric, show that the above equation can be written in terms of cross product as defined below (rotation vector $\mathbf{\omega}$):

$$\mathbf{v} = \mathbf{\omega}\,(t) \times \mathbf{x}t + \mathbf{C}(t)$$

3.11. (a) A particular flow field is given as $v_x = ax$, $v_y = -ay$. A pollutant in the following fluid has the following concentration:

$$y > 0;\ a \ \& \ b \text{ are constants}$$

Does the concentration for a particular fluid element change with time?

(*Hint*: Evaluate the material derivative of c.)

(b) For the velocity field $\mathbf{v} = v_x(y)\mathbf{e}_x$, evaluate material and the two convected derivatives of **D**.

3.12 (a) Based on the following description, derive a components' balance. Consider a material system consisting of i components. Let ρ_i be the density of each component at a material point. The sum of all the components' densities is equal to the average density of the material at a material point, so that $\sum \rho_i = \rho$. A component mass balance can be written for the amount of component i in a region D^t of material. The rate of change of amount of component in D^t can be said to be equal to the amount transferred from S^t and the amount reacted or produced within the region. Hence, similar to the case of other balances, we have a surface term and a volumetric term. Assume that the surface term can be estimated based on the hypothesis of a mass flux of component i, $\mathbf{j}_i$. The velocity of a material point is $\mathbf{v}$ or $\dot{\mathbf{x}}$, as earlier.

(b) Show that when you add all the components' balances, the resulting equation is the mass balance.

3.13. Start from the energy balance and show that it can be reduced to the conduction equation,

$$\frac{\partial \theta}{\partial t} = \frac{\kappa}{\rho C} \frac{\partial^2 \theta}{\partial x^2}$$

The following assumptions are required to be made in the above derivation:

(a) Internal energy is a function of temperature only ($\varepsilon = \varepsilon_R + C\theta$, where C is the heat capacity of the material and ε_R is a constant).

(b) Heat flux is described by Fourier's law ($\mathbf{q} = -\kappa \operatorname{grad} \theta$).

(c) Volumetric heat source can be neglected ($r = \mathbf{0}$).

(d) The material does not undergo any deformation ($\mathbf{u} = 0$).

3.14. Power due to contact forces appears as $\boldsymbol{\sigma} : \mathbf{D}$ or tr ($\boldsymbol{\sigma} \cdot \mathbf{D}$). Evaluate the same power in terms of first Piola-Kirchoff stresses.

(*Hint:* Substitute for $\boldsymbol{\sigma}$ and $\mathbf{D}$ in terms of first Piola-Kirchhoff stresses and deformation gradient, respectively).

3.15 (a) Consider an ideal gas in the absence of heat transfer and generation ($h = r = 0$). Assume the ideal gas to be an ideal fluid (inviscid fluid, $\boldsymbol{\sigma} = -p\mathbf{I}$). Simplify the energy balance.

(b) Assume that internal energy of an ideal gas to be $\varepsilon = \varepsilon_R + C_v \theta$, where C_v is specific heat at constant volume. Based on the energy balance, show that $\theta = k\rho^{R/C_v}$, where k is a constant. Therefore, this result is equivalent to, for an ideal gas with adiabatic processes, $PV^\gamma = \text{constant}$.

3.16. Show that balance of linear momentum can be written in the reference configuration as

$$\rho_0 \dot{\mathbf{v}} = \text{Div } \mathbf{S} + \rho_0 \mathbf{b}$$

3.17. Using the steps followed in development of Eqn. (3.4.6), show that an objective tensor **T** transforms as follows :

$$\mathbf{T}^* = \mathbf{Q}\mathbf{T}\mathbf{Q}^T.$$

3.18. Show that the velocity (*v*), spin tensor (**W**) and the stretching tensor (**D**) transform as follows, with respect to an orthogonal transformation:

$$\mathbf{v}^* = \mathbf{Q}\mathbf{v} + \dot{\mathbf{Q}}\mathbf{Q}^T\mathbf{x}^*$$

$$\mathbf{W}^* = \mathbf{Q}\mathbf{W}\mathbf{Q}^T + \dot{\mathbf{Q}}\mathbf{Q}^T$$

$$\mathbf{D}^* = \mathbf{Q}\mathbf{D}\mathbf{Q}^T$$

(*Hint:* Follow Example 3.4.1)

3.19. If **T** is a frame invariant tensor, show that $\overset{\Delta}{\mathbf{T}}$ and $\overset{\nabla}{\mathbf{T}}$ are also frame indifferent.

3.20. Show that expression for stress tensor given in Eqn. (3.4.26) can be written as that given in Eqn. (3.4.27).

■■■

CHAPTER

 Linear Mechanical Models
of Material Deformation

यथा सोम्य एकेन मृतिपण्डेन सर्व मृन्मयं विज्ञातं स्यात् वाचारम्भणं विकारो नामधेयं
मृत्तिकेत्येव सत्यम् ।   (तैत्तिरीयोपनिषत्)
Yathā somya ekena mṛitpindena sarvaṃ mrinmayam vigyataṃ syāt
vācarambhanaṃ vikāro naamdhyum mṛttiketyeva satyaṃ
                                                                                                          (Taitttiriyopaniṣad)

O somya, just as all clay products are known by knowing a lump of clay,
clay alone is the truth and all clay products are mere verbal modifications
of clay.

4.1 INTRODUCTION

In Chapter 3, we broadly outlined the kinematics of material deformation, the
balance laws and the need for constitutive relations to solve any boundary
value/initial value problem associated with material deformation. All the
concepts have been developed in very general way assuming that the material
can undergo any arbitrary deformation. Even though the focus of this book
will continue to be analysis of material deformation from this general
perspective, it would be worthwhile recapitulating models of material
deformation, that are very popular in engineering. This is necessary to see
how these popular models fit within the broader framework that is promised
by the description of continuum mechanics outlined in Chapter 3.

When we examine engineering practice, we observe that most of the early
popular material models have been *linear* models. Linearity refers to the nature
of relationship between a chosen output variable to a chosen input variable.
For example, we can think of stress as an output variable and strain or strain
rate as input variables. A linear relation implies that a scaled change in input

variable will change the output variable also by the same scaling. In other words, the output of a linear model from a combination of inputs is the same as the combination of outputs from individual inputs.

The linear models have been visualized as simple extensions of empirical models of material deformation observed in simple experiments. For example, most solids were visualized to behave like Hookean solids, where a linear relationship is assumed to exist between the stresses and strains. Similarly, most fluids were visualized to behave like Newtonian liquids, where a linear relationship is assumed to exist between stresses and strain rates. On the other hand, linearity in linear viscoelastic models implies a linear combination of elastic and viscous responses.

While the general deformation behaviour of many materials is complex than is visualized by these simplistic models, the ability of these simplistic models to capture many observed experimental response of various common engineering materials cannot be underestimated. The simplistic linear models have also been popular in design for many reasons, some of which are outlined below:

(i) Linear models are easy to use and deal with a minimum number of material parameters that will characterize a mechanical response. For example, it is easy for a designer to visualize that there is a single parameter called modulus that will characterize the way a material will deform when we applied loads. The designers would like to interpret any other complicated behavior; such as the response at large deformations, time dependent deformations etc., in terms of the way these factors influence a chosen single parameter like modulus or viscosity.

(ii) Most materials are made to operate within a linear range in design. Non-linear response in materials, especially in solids, usually signifies the onset of damage and deterioration in the material, which is avoided for most normal operations in design. Similarly in transport of fluids, assumption of linear behaviour in fluids with a single viscosity parameter is usually sufficient to estimate the pumping requirements.

It is useful for us to recollect the popular linear models that are commonly used in engineering, so that it will be easy to interpret complex behaviour in terms of understanding obtained from the simplistic linear models. Hence, the popular linear models for both solidlike and fluidlike materials will be reviewed in the current chapter.

4.2 LINEAR ELASTIC SOLID MODELS

The most popular *linear elastic solid model* is the elastic model that is defined in *Hookean elasticity*. The linear elastic models are visualized as extensions of the experimental observations that are attributed to Hooke, Euler and Young.

Hooke observed that the load in any solid is linearly related to the elongation. Euler and Young later visualised the existence of a material property that is a part of this linear relation.

We learnt in Chapter 3, that a characterization of load-elongation relation is better expressed as a stress-strain relation. Elasticity implies that there is a unique strain associated with any applied stress, and a unique stress corresponding to an applied strain. This would normally imply that the material will retrace the stress-strain trajectory when it is unloaded, as shown in Fig. 4.1 as curves (a) and (b). It is useful to note that a material may be elastic but the response need not be linear, as is suggested by curve (b). Such materials are called *non-linear elastic* materials and will be described in Chapter 6. There are *inelastic* materials (example of elastic-plastic material given in curve (c) in Fig. 4.1), which do not retrace their stress-strain trajectory, when they are unloaded. The inability of the material to retrace its loading path is primarily due to dissipative mechanisms such as *plasticity*, in the material. These phenomena will also be dealt with in Chapter 6. The stress-strain relationships of solids which are linear and allow for retracing while loading and unloading will be discussed in this section. These relationships are called linear elastic solid models of material behaviour. They are also said to be Hookean elastic models and linear elastic models.

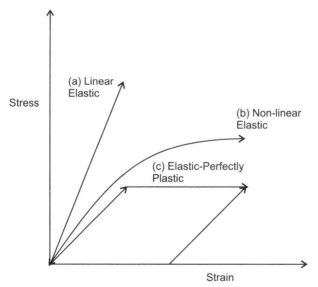

Fig. 4.1 Schematic representation of Material response of various material

In modelling of materials, it is usually understood that the stresses and strains are both referred to the same configuration of the body. In other words, both of them should be described either in the reference configuration or in the deformed configuration. For example, the balance laws or equilibrium equations that describe the stresses are written in the deformed configuration (Eqn. (3.3.21)). Therefore, the strains (such as Eqn. (3.2.61)) should also be expressed in the deformed configuration. Similar stresses and strain measures were defined in Chapter 3 in the reference configuration. The choice of different stress and strain measures is discussed in Chapter 6 on Non-linear models. In linear elasticity, no distinction exists among the different stress measures (and strain measures) due to the assumption of small strains.

4.2.1 Small Strain Assumption of Linear Elasticity

In linear elasticity, it is assumed that the deformations are infinitesimally small and a mathematical relation between the stress and strain can be described. In usual description of linear elasticity, stresses are defined in terms of deformed configuration, while the strains are defined in terms of reference configuration. This is possible because all the alternate measures can be shown to reduce to infinitesimal strain tensor **e** defined in Eqn. (3.3.9).

Having stated that the stress σ and the strain **e** at a point can be related to each other, the stress can be expressed in terms of the strain as

$$\sigma = f(\mathbf{e}) \tag{4.2.1(a)}$$

or
$$\sigma_{ij}(e_{kl}) = \sigma_{ij}^{0} + C_{ijkl}e_{kl} + O\left(e_{kl}^{2}\right) \tag{4.2.1(b)}$$

where Eqn. (4.2.1(b)) is an expansion of Eqn. (4.2.1(a)) in index notation.

Assuming that the reference stress σ_{ij}^{0} to be zero and ignoring the quadratic and higher order terms $\left(O\left(e_{kl}^{2}\right)\right)$, we find that there is a linear relation between the stress σ_{ij} and strain e_{kl}. Constants C_{ijkl} (set of 81 constants) characterize the linear relation and they are independent of the state of stress or strain in the body. These constants are in fact material parameters describing the mechanical response of a material. In the Section 4.2.2, we will examine the specific forms of these constants for various classes of materials that are common in solid mechanics. In this chapter, the useful expressions associated with the common elastic materials, are documented.

4.2.2 Classes of Elastic Constants

The balance of angular momentum introduced in Eqn. (3.3.25) tells us that the Cauchy stress tensor will always be symmetric. The symmetric form of strain-displacement relations shows that the strain tensors are also symmetric. The symmetry of both stress and strain tensors imply that there are only *six* independent stresses and strains that may need to be related to each other through a constitutive relation. Hence, the most general form of the Hooke's law, which relates these stresses to the strains, is of the form

$$\sigma_\alpha = C_{\alpha\beta} e_\beta; \qquad \alpha, \beta = 1, 2, ..., 6, \qquad (4.2.2)$$

where

$$\sigma_1 = \sigma_{11}, \ \sigma_2 = \sigma_{22}, \ \sigma_3 = \sigma_{33}, \ \sigma_4 = \sigma_{12}, \ \sigma_5 = \sigma_{23}, \ \sigma_6 = \sigma_{13};$$
$$e_1 = e_{11}, \ e_2 = e_{22}, \ e_3 = e_{33}, \ e_4 = e_{12}, \ e_5 = e_{23}, \ e_6 = e_{13}. \qquad (4.2.3)$$

Equation (4.2.2) implies that there are 36 constants which relate the stresses to strains.

4.2.2.1 General Anisotropic Linear Elastic Solid

As discussed in Section 3.3.6.1, every solid will have an intrinsic scalar *energy* density field, which is in the form of an elastic strain energy density $W(e_\alpha)$. This strain energy density can be expressed in terms of Helmholtz free energy as follows,

$$W = \rho_o \psi. \qquad (4.2.4)$$

The stresses can be defined as (analogous to Eqn. (3.4.25)),

$$\sigma_\alpha = \frac{\partial W}{\partial e_\alpha}; \qquad \alpha = 1, 2, ..., 6. \qquad (4.2.5)$$

Equation (4.2.5) will be dicussed in more detail in Section 6.2. Substituting Eqn. (4.2.5) in Eqn. (4.2.2) and differentiating, we get

$$\frac{\partial^2 W}{\partial e_\alpha \partial e_\beta} = C_{\alpha\beta} \qquad (4.2.6)$$

Since Eqn. (4.2.6) indicates symmetry in the differentiation of W, we can easily see that

$$C_{\alpha\beta} = C_{\beta\alpha}. \qquad (4.2.7)$$

Equation (4.2.7) states that there is symmetry in the linear stress-strain relationship in an elastic body. Taking this symmetry into account, if we count the number of independent constants in an elastic body, we will find that there are 21 independent constants in an elastic body. If all these constants are distinct and unrelated, the body is called as a *general anisotropic elastic* body. The constants for a general anisotropic elastic body can be represented in a matrix form as shown in Eqn. (4.2.8) below:

$$
\mathbf{C} = \begin{pmatrix}
C_{11} & C_{12} & C_{13} & \cdot & \cdot & C_{16} \\
 & C_{22} & C_{23} & \cdot & \cdot & \cdot \\
 & & C_{33} & \cdot & \cdot & \cdot \\
 & & & C_{44} & \cdot & \cdot \\
 & Sym & & & C_{55} & C_{56} \\
 & & & & & C_{66}
\end{pmatrix} \qquad (4.2.8)
$$

4.2.2.2 Materials with Single Plane of Elastic Symmetry

In many elastic solids, a symmetry in mechanical response may exist with respect to a transformations about a plane. Consider a transformation (x^*, y^*, z^*) from (x, y, z). A body with one plane of elastic symmetry $(xy$ plane) is characterized by the transformation $x^* = x$, $y^* = y$, $z^* = -z$. It is possible to prove (see Exercise 4.1) that for such materials the following relations are true:

$$C_{15} = C_{16} = C_{25} = C_{26} = C_{35} = C_{36} = C_{45} = C_{46} = 0. \qquad (4.2.9(a))$$

Hence, materials with one plane of elastic symmetry will have 13 independent constants. The matrix of constants for such a material can be expressed as

$$\mathbf{C} = \begin{pmatrix} C_{11} & C_{12} & C_{13} & C_{14} & 0 & 0 \\ & C_{22} & C_{23} & C_{24} & 0 & 0 \\ & & C_{33} & C_{34} & 0 & 0 \\ & & & C_{44} & 0 & 0 \\ & Sym & & & C_{55} & C_{56} \\ & & & & & C_{66} \end{pmatrix} \qquad (4.2.9(b))$$

4.2.2.3 Materials with Two Planes of Elastic Symmetry

In some materials, there may exist two planes of elastic symmetry, (say xy plane and xz plane). This can be simulated by superimposing a transformation $x^* = x$, $z^* = z$, $y^* = -y$ on the material with a single plane of symmetry outlined in Eqn. (4.2.9). It is possible to prove that for such material the following constants will not exist as well. Therefore, we have

$$C_{14} = C_{24} = C_{34} = C_{56} = 0. \qquad (4.2.10)$$

Materials with two planes of symmetry will have 9 constants and are generally termed as *orthotropic materials*. The matrix of constants for such materials can be expressed as

$$\mathbf{C} = \begin{pmatrix} C_{11} & C_{12} & C_{13} & 0 & 0 & 0 \\ & C_{11} & C_{23} & 0 & 0 & 0 \\ & & C_{33} & 0 & 0 & 0 \\ & & & C_{44} & 0 & 0 \\ & Sym & & & C_{55} & 0 \\ & & & & & C_{66} \end{pmatrix} \qquad (4.2.11)$$

4.2.2.4 Materials with Symmetry about an Axis of Rotation

Many times we also encounter materials which have an axis of symmetry. Let the axis of symmetry of such materials be the z axis. These materials show an independence or property variation with rotations about a plane perpendicular to the axis of symmetry (z axis) and are called as *transversely isotropic materials*. In such materials, many of the coefficients in the array given in (Eqn. (4.2.11),

will be related ($C_{11} = C_{22}$, $C_{13} = C_{23}$, $C_{55} = C_{66}$ and $C_{44} = \dfrac{1}{2}(C_{11} - C_{12})$) and the

matrix of elastic constants is given by

$$
\mathbf{C} = \begin{pmatrix}
C_{11} & C_{12} & C_{13} & 0 & 0 & 0 \\
 & C_{11} & C_{23} & 0 & 0 & 0 \\
 & & C_{33} & 0 & 0 & 0 \\
 & & & C_{44} & 0 & 0 \\
 & Sym & & & C_{55} & 0 \\
 & & & & & C_{66}
\end{pmatrix}
\qquad (4.2.12)
$$

Therefore, five independent coefficients of $C_{\alpha\beta}$ are needed to be described for a transversely isotropic elastic solid. It is further possible to show that the coefficients given in Eqn. (4.2.12) are related to each other through the engineering coefficients like Young's moduli, shear moduli, and Poisson's ratios so that we have the following relations

$$
C_{11} = \frac{\left(1 - n v_{zx}^2\right) E_x}{AB}, \quad C_{12} = \frac{\left(v_{xy} + n v_{zx}^2\right) E_x}{AB},
$$

$$
C_{13} = \frac{v_{zx} E_x}{B}, \qquad C_{33} = \frac{\left(1 - v_{xy}\right) E_z}{B}, \quad C_{55} = G_{xz} = G_{yz} \qquad (4.2.13)
$$

where $A = 1 + v_{xy}$, $B = 1 - v_{xy} - 2 n v_{zx}^2$, $n = \dfrac{E_x}{E_z}$

Additionally, we have the following relationships are used in an engineering description of these materials,

$$
G_{xy} = \frac{E_x}{2\left(1 + v_{xy}\right)}, \text{ and } v_{xz} = n v_{zx} \qquad (4.2.14)
$$

From the description of Eqn. (4.2.13), we find that there are only five independent constants that are necessary to describe a transversely isotropic material, and they are E_x, E_z, v_{xy}, v_{zx}, and G_{xz}. These constants are defined from an engineering standpoint, *i.e.*, from the standpoint of measurements that can be actually made on test specimens. The first two constants represent the Young's modulii in the x and z directions respectively and are defined as the ratio of the tensile (or compressive) stress and tensile (or compressive) strain) experienced in a direct tension (or compression) test. The next two constants are the Poisson's ratios measured in $x - y$ and $x - z$ planes respectively. They are the absolute value of the lateral strain and longitudinal strains experienced by the material in a direct tension test. The constant n is the ratio of the moduli in the two directions and represents the degree of anisotropy that exists in the material. G_{xz} is the shear modulus measured in the $x - z$ plane.

G_{xy} is the shear modulus measured in the $x - y$ plane and is related to Young's modulus and the Poisson's ratio, as indicated in Eqn. (4.2.14). The relations given in Eqns. (4.2.13) and (4.2.14) describe the linear elastic constants of a transversely isotropic material are widely used in composite materials and in geologic layered materials like oil shale etc.

4.2.2.5 Isotropic Materials

Materials whose elastic properties do not change with any direction, are called as *isotropic materials*. Very often, the assumption of isotropy goes along with another assumption of *homogeneity*, i.e., the material properties are assumed to be independent of the position of the material point within the material. Metals, ceramics and polymers are generally considered to be linear isotropic materials for small strain applications.

For an isotropic body, it is possible to prove that the principal strain directions and the principal stress directions coincide with each other. For such materials, it is possible to reduce the number of independent elastic constants to just *two* constants. These two constants are normally referred to as *Lame*'s parameters and are designated by the symbols λ and μ. These parameters can be shown to be related to the general elastic coefficients $C_{\alpha\beta}$ that we have been discussing in Eqns. (4.2.8) – (4.2.12). It is possible to show that for an isotropic elastic body, the stresses and strains are related to each other through the expression

$$\sigma_{ij} = \lambda e \delta_{ij} + 2\mu e_{ij}, \quad i, j = 1, 2, 3 \tag{4.2.15}$$

where $e = e_{11} + e_{22} + e_{33}$ is the volumetric strain due to small deformation. It is easy to see that Eqn. (4.2.15) represents a set of six independent equations which are given by

$$\sigma_{11} = \lambda e + 2\mu e_{11}, \quad \sigma_{22} = \lambda e + 2\mu e_{22}, \quad \sigma_{33} = \lambda e + 2\mu e_{33},$$

$$\sigma_{12} = 2\mu e_{12}, \qquad \sigma_{23} = 2\mu e_{23}, \qquad \sigma_{13} = 2\mu e_{13} \tag{4.2.16}$$

Equation (4.2.16) can be inverted to obtain the strain-stress relations as

$$e_{11} = \frac{1}{E}\left(\sigma_{11} - v\left(\sigma_{22} + \sigma_{33}\right)\right), \qquad e_{22} = \frac{1}{E}\left(\sigma_{22} - v\left(\sigma_{11} + \sigma_{33}\right)\right),$$

$$e_{33} = \frac{1}{E}\left(\sigma_{33} - v\left(\sigma_{11} + \sigma_{22}\right)\right),$$

$$e_{12} = \frac{1}{2G}\sigma_{12} = \frac{1+v}{E}\sigma_{12}, \qquad e_{13} = \frac{1}{2G}\sigma_{13} = \frac{1+v}{E}\sigma_{13},$$

$$e_{23} = \frac{1}{2G}\sigma_{23} = \frac{1+v}{E}\sigma_{23}, \tag{4.2.17}$$

where $G = \mu$, $E = \dfrac{G(3\lambda + 2G)}{\lambda + G}$, and $v = \dfrac{\lambda}{2(\lambda + G)}$.

From an engineering standpoint, G is called as the *shear modulus*, which represents the linear relationship between the shear stresses and the shear strains. E is called the *Young's modulus* and represents the linear relationship between axial stresses and strains. v is the *Poisson's ratio* and represents the ratio of the lateral strain to the longitudinal strain in a uniaxial test. For an isotropic elastic body, it is also possible to define the following additional relations

$$G = \frac{E}{2(1+v)} \quad \text{and} \quad \kappa = \frac{E}{2(1-2G)}, \tag{4.2.18}$$

where κ is the *bulk modulus* of the material and represents the ratio between the volumetric stresses and strains in a material.

It is important to note that even when we use the engineering definitions *viz.*, E, v, G and κ, it is sufficient that we define two of the material properties independently for an isotropic elastic material. All other quantities can be seen to be related to these two independently defined quantities. Very often the Young's modulus E and the Poisson ratio v are defined independently for the material. Further, when even simpler description of the elastic response of an isotropic elastic solid is desired, only the Young's modulus E is used to describe the stress-strain response of an isotropic elastic solid. We will also observe the use a single modulus parameter in the development of *viscoelastic response* of materials in Section 4.4.

4.3 LINEAR VISCOUS FLUID MODELS

Similar to the uniaxial tension experiment on solid materials, fluid behaviour is described based on a shear experiment. As shown in Fig. 4.2, fluid flow between two parallel plates is considered. One of the plate moves with a constant velocity, while the other remains stationary.

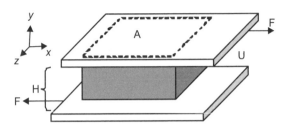

Fig. 4.2 Sketch describing the simple shear experiments on fluids

Based on experiments of this type, stress-strain rate behaviour is used to classify different fluids. Fig. 4.3 shows some examples of stress *vs* strain rate behaviour in fluids. Curve (a) represents a linear relationship between stress and strain rate, while curve (b) represents non-linear relationship between stress and strain rate. Fluids of the type represented by curve (b) are discussed in Chapter 5. Here, we will describe the models which describe the linear behaviour between stress and strain rate. Curve (c) represents a special class of materials for which the stress is independent of the strain rate. The relationship of this nature is also described in this section in the following discussion.

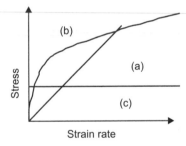

Fig. 4.3 Stress *Vs* strain rate behaviour for different fluids

4.3.1 General Anisotropic Viscous Fluid

For viscous fluids, it is assumed that the stresses in the material are dependent only on the stretching (popularly called the strain rate). The two most common models used for fluids are *ideal* (also called *perfect* or *inviscid*) fluid and *Newtonian* fluid. In this section, these models are introduced. Based on the assumptions for viscous fluid given above, the constitutive relation for a viscous fluid will be of the following form,

$$\sigma = f(\mathbf{D}) \tag{4.3.1(a)}$$

or

$$\sigma_{ij} = \sigma_{ij}^{0} + C_{ijkl}D_{kl} + O\left(D_{kl}^{2}\right), \tag{4.3.1(b)}$$

where Eqn. (4.3.1(b)) is an expansion of Eqn. (4.3.1(a)) in index notation.

When a fluid is at rest, the stress tensor is given by hydrostatic pressure. D_{kl} will be zero when fluid is at rest, as introduced in Eqn. (3.2.46). To be consistent with this requirement, σ_{ij}^0 is taken to be

$$\sigma_{ij}^0 = -p\delta_{ij} \qquad (4.3.2)$$

In the current chapter, we will only examine the linear viscous fluid. The constitutive relation for linear viscous fluid, which is a specific case of Eqn. (4.3.1), is as follows:

$$\sigma_{ij} = -p\delta_{ij} + C_{ijkl}D_{kl} \qquad (4.3.3)$$

Equation (4.3.3) suggests that there is a linear relation between stress and stretching (strain rate).

4.3.2 Isotropic Viscous Fluid

Analogous to the discussion followed for linear elastic solids, the number of constants represented by C_{ijkl} can also be reduced. For an isotropic linear viscous fluid, the number of independent constants can be shown to be two. Hence, for linear viscous isotropic fluid can be described by the following relation:

$$\sigma_{ij} = -p\delta_{ij} + \lambda I_D \delta_{ij} + 2\mu D_{ij} , \qquad (4.3.4)$$

where λ is *bulk* or *dilatational viscosity*. μ is called as *dynamic* or *shear viscosity*, popularly known just as *viscosity*. The fluids represented by Eqn. (4.3.4) are called Newtonian fluids. More commonly, we encounter incompressible ($I_D = 0$) Newtonian fluids described by the following constitutive relation

$$\sigma_{ij} = -p\delta_{ij} + 2\mu D_{ij} . \qquad (4.3.5)$$

Based on Eqn. (3.4.37), the deviatoric stress for incompressible Newtonian fluids is given as

$$\tau_{ij} = 2\mu D_{ij} . \qquad (4.3.6)$$

Air, water and oils are common examples of the linear visocous fluids. Another important class of fluids encountered in aerodynamics are called *inviscid*, *Euler* or *ideal* fluids, in which case the stress is given by

$$\sigma_{ij} = -p\delta_{ij} . \qquad (4.3.7)$$

Equation (4.3.6) tells us that the viscosity, μ is the most important material property that is characterized for a incompressible linear viscous fluid. Hence, other fluids are also very often characterized using a single viscosity parameter, μ. Use of μ as a parameter will be illustrated in modelling of viscoelastic materials, described in Section 4.4.

4.4 VISCOELASTIC MODELS

As seen in Section 4.2, for elastic materials stress was stated to be a function of strain only. Similarly, in Section 4.3, the stress was stated to be a function of strain rate only. However, for several materials, the stress has been observed to be function of strain, strain rate as well as the stress rate itself. These materials are called viscoelastic materials. Analogous to the most general equations for an elastic material (Eqn. (4.2.1)) and for a viscous material (Eqn. (4.3.1)), it is possible to state general equations for viscoelastic materials. Initially, for ease in understanding, we will describe the viscoelastic material response as combinations of elastic and viscous responses in one dimension. In Section 5.4, we discuss non-linear models of viscoelasticity.

Both in case of linear elastic and linear viscous fluids, the material properties (E and μ) have been assumed to be independent of time. This is not the case for description of material behaviour in case of metals at high temperature, paints, asphalt, blood, polymer melts, gums, resins and foodstuffs. We know that metals that are held at constant load at high temperatures exhibit continuous deformation with time. This phenomenon is generally termed as *creep*. A polymer sample when subjected to a constant stretch, requires less and less load to maintain the stretch. This phenomenon is called *stress relaxation*. In both these cases, we observe a time dependent nature of the modulus (defined as stress per unit strain). Time dependent behaviour is also observed when many of the above materials are subjected to time dependent (usually *oscillatory*) stress or strain. We notice that in all the above cases, the material seems to have elements of both solidlike and fluidlike behaviour. Hence, it will be useful to combine both these behaviours using parameters such as (elastic) modulus and viscosity. These materials and the associated models are referred to as *viscoelastic*.

As was discussed in case of solid like material, load-extension behaviour popularly implies uniaxial extension. Similarly, in case of fluidlike materials, load-strain rate behaviour popularly implies simple shear flow. Viscoelastic behaviour may be observed either in extension or in shear. Regardless of the mode of observation, same set of simplistic models can be used to describe the viscoelastic behaviour. Additionally, simplistic models are usually one dimensional. Hence, it will be sufficient to use uni-dimensional measures such s for stress and **e** for strain.

Simplistic models describe the viscoelastic behaviour only for small strains. It can be shown that for small strains, **D** is equivalent to $\dot{e}$, infinitesimal strain rate tensor. Therefore, in this section we will use $\dot{e}$ to denote strain rate for a one dimensional observation. Viscoelastic behaviour at small strains is referred to as the *linear viscoelastic* behaviour, as it can be modelled using a linear combination of linear elastic and linear viscous behaviour. The linearity of combination implies a scaled change in input variable (stress, strain or strain rate) will change the output variable (strain or strain rate and stress) also by the same scaling.

4.4.1 Useful Definitions for Description of Viscoelastic Behaviour

As was discussed in Section 4.4, creep, stress relaxation and oscillatory observations are useful in describing viscoelastic behaviour of any material. It was highlighted that parameters such as modulus and viscosity are time dependent for viscoelastic materials. This time dependence of material properties is captured through *material functions*. These material functions are outlined below.

4.4.1.1 Creep Compliance and Relaxation Modulus

In creep, a constant stress (σ^0) is applied, and a material function called *creep compliance,* (J_c) is defined as

$$J_c(t) = \frac{e(t)}{\sigma^0}.$$ (4.4.1)

We notice from Eqn. (4.4.1) that the creep compliance is a function of time, since strain is a function of time. In stress relaxation, a constant strain (e^0) is applied. In this case, a material function called *relaxation modulus* (E_r) is defined as

$$E_r(t) = \frac{\sigma(t)}{e^0}.$$ (4.4.2)

Relaxation modulus thus defined is also a function of time, since stress is a function of time.

We understand that viscoelastic behaviour has an energy storing function (characterized by the elastic property) and an energy dissipating function (characterized by the viscous property). It is often useful to characterize these two contributions using a simple test. This is done by subjecting a viscoelastic material to an oscillatory loading. This oscillatory loading (dynamic loading) is usually a sinusoidal loading and the response of the material is studied with respect to this loading.

4.4.1.2 Phase Lag, Storage Modulus and Loss Modulus

We recollect the following relationship one-dimensional relationship between stresses in an elastic solid (σ_e) and strains:

$$\sigma_e(t) = Ee(t). \tag{4.4.3}$$

Similarly, we recollect the following one-dimensional relationship between the stresses in a viscous fluid (σ_v) and the strain rates:

$$\sigma_v = \mu\dot{e}(t). \tag{4.4.4}$$

Assuming that the $e(t)$ and $\dot{e}(t)$ in Eqn. (4.4.3) and Eqn. (4.4.4) are in the form of a sinusoidal loading, we can assume the following expressions for strain and strain rate in the material:

$$e(t) = e^0 \sin \omega t; \quad \dot{e}(t) = \dot{e}^0 \cos \omega t, \tag{4.4.5(a)}$$

where e^0 and $\dot{e}^0$ are the maximum amplitude of strain and strain rates assumed in the sinusoidal loading. In Eqn. (4.4.5), we have deliberately chosen a sine function to represent the strain and a cosine function to represent the strain rates. This is to indicate that the strains and strain rates are always out of phase with respect to each other. If both these loadings are applied to the same material, it is easy to see that

$$\dot{e}^0 = \omega e^0. \tag{4.4.5(b)}$$

Substituting for strains from Eqn. (4.4.5(a)) in Eqn. (4.4.3) and for strain rates from Eqn. (4.4.5(a)) in Eqn. (4.4.4), we get the following expressions for σ_e and σ_v as

$$\sigma_e(t) = Ee^0 \sin \omega t, \quad \sigma_v(t) = \mu\dot{e}^0 \cos \omega t. \tag{4.4.6}$$

Comparing the expressions for stress, strain and strain rate from Eqn. (4.4.5) and Eqn. (4.4.6), we observe that a sinusoidal strain rate is always *in phase* with the associated viscous stress. However, a sinusoidal strain is always *out of phase* with the associated viscous stress. Similarly, a sinusoidal strain will be in phase with the associated elastic stress and a sinusoidal strain rate will be out of phase with the associated elastic stress. These concepts are clearly illustrated in Fig. 4.4. In phase and out of phase can also be written in terms of phase lags being 0 and 90°, respectively.

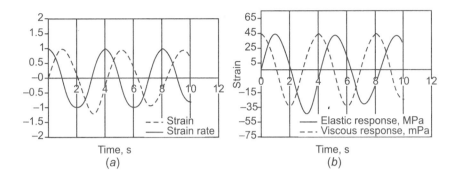

Fig. 4.4 (a) Sinusoidal (or dynamic) loading (b) Response of elastic and viscous materials to the loading given in (a)

For a viscoelastic material, the phase lag between the stress and the strain (or strain rate) has been observed to be between 0 and 90°. Additionally, viscosity and modulus for a viscoelastic material are functions of frequency. Given the applications of strain or strain rate (Eqn. (4.4.5)), the general stress response can be written as

$$\sigma(t) = \mu^*(\omega)\dot{e}^0 \cos(\omega t + \delta)$$
$$\sigma(t) = E^*(\omega)e^0 \sin(\omega t + \xi). \tag{4.4.7}$$

We observe from Eqn. (4.4.7) that $\mu^*(\omega)$ and $E^*(\omega)$ are material functions signifying the viscosity and modulus of the viscoelastic material. In Eqn. (4.4.7), $\delta = 0$ and $\xi = 90°$ imply viscous response and $\delta = 90°$ and $\xi = 0$ imply elastic response. Equation (4.4.7) can be rewritten in terms of two terms, in phase and out of phase with respect to strain or strain rate, *viz.*,

$$\sigma(t) = \dot{e}^0 \mu^*(\omega)\cos\delta \cos\omega t - \dot{e}^0 \mu^*(\omega)\sin\delta \sin\omega t$$
$$\sigma(t) = e^0 E^*(\omega)\cos\xi \sin\omega t + e^0 E^*(\omega)\sin\xi \cos\omega t \tag{4.4.8}$$

From Eqn. (4.4.8), we can define

$$\mu'(\omega) = \mu^*(\omega)\cos\delta$$
$$\mu''(\omega) = \mu^*(\omega)\sin\delta, \tag{4.4.9}$$

and

$$E'(\omega) = E^*(\omega)\cos\xi$$
$$E''(\omega) = E^*(\omega)\sin\xi. \tag{4.4.10}$$

Substituting Eqns. (4.4.9) and (4.4.10) in Eqn. (4.4.8), we obtain

$$\sigma(t) = \dot{e}^0 \mu'(\omega)\cos\omega t - \dot{e}^0 \mu''(\omega)\sin\omega t$$
$$\sigma(t) = e^0 E'(\omega)\sin\omega t + e^0 E''(\omega)\cos\omega t. \tag{4.4.11}$$

Different interpretations are given to the material functions $\mu'(\omega)$, $\mu''(\omega)$, $E'(\omega)$ and $E''(\omega)$, as they characterize the material response that is viscous or elastic. For example, E' is referred to as the *storage modulus* as it characterizes the elasticlike (energy storing) response of a viscoelastic material. E'' characterizes the viscouslike (energy dissipating) behaviour and is called the *loss modulus*. However, both are important in overall characterization of a viscoelastic material and therefore, sometimes combination of them is expressed in the form of a *complex modulus*:

$$E^* = E' + iE''. \tag{4.4.12}$$

Similarly, *complex viscosity* is expressed as

$$\mu^* = \mu' - i\mu''. \tag{4.4.13}$$

We note that the ratios of moduli and viscosities can be written in terms of phase lags

$$\tan\delta(\omega) = \frac{\mu'(\omega)}{\mu''(\omega)}, \ \tan\xi(\omega) = \frac{E'(\omega)}{E''(\omega)}. \tag{4.4.14}$$

In this section, we have defined the creep compliance, stress relaxation modulus and the storage and loss moduli for a viscoelastic material. Similarly, complex viscosity was also defined. These concepts will be used to examine the simplistic models of viscoelasticity in Section 4.4.2.

4.4.2 Simplistic Models of Viscoelasticity

In Sections 4.2 and 4.3, we described the simplistic models for a linear elastic solid and a linear viscous fluid. Simplistic models of viscoelasticity can be visualized as combinations of these two models. The visualization of the elastic and viscous responses of a viscoelastic material can be done in several ways. It is possible to visualize the kinetic measures of the material response such as total stresses can be an additive combination of the viscous and elastic behaviour or the material. The *Kelvin-Voigt* model that is popular in the literature works with such a visualization of the total response. It is also possible to visualize the kinematic measures of the material, such as total strain rates to be an additive combination of the strain rates of viscous and elastic elements of the material response. The *Maxwell model* of viscoelasticity, that is popular in literature is developed with such a visualization in mind. It should be noted that parameters such as modulus and viscosity of Kelvin and Maxwell models may not have any correlation with modulus of elasticity and shear viscosity. This will be discussed in greater detail in Section 4.4.2.1.

It is also possible to visualize the total mechanical response of the viscoelastic material as many combinations of these two responses. However, our attempt here will be observe one such model in great detail, even while outlining the possibilities that are possible with other similar models, which are popularly used to describe viscoelastic behaviour.

4.4.2.1 Maxwell Model

Governing Equation for the Maxwell Model

One of the ways in which the governing equation of the *Maxwell model* can be understood is described below. As indicated earlier, the Maxwell model visualizes that the total kinematics in the material can be arrived at by adding the kinematics associated with elastic and the viscous part of the material. It is assumed that the elastic and viscous strain responses, under the action of same stress, can be added to obtain the overall response of a material. Denoting the elastic (recoverable) strain in the material as e_e and the viscous (irrecoverable) strain in the material as e_v, we have

$$e = e_e + e_v,\qquad(4.4.15)$$

where, e is the strain response of the viscoelastic material. Taking a simple time derivative of Eqn. (4.4.15), we get

$$\dot{e} = \dot{e}_e + \dot{e}_v,\qquad(4.4.16)$$

where $\dot{e}$ is the strain rate response of the viscoelastic material. Noting that both the above responses arise due to the action of the same stress σ (which is the same as σ_e and σ_v introduced in Eqns. (4.4.3) and (4.4.4)) in a material, we get

$$\dot{e} = \frac{1}{E}\dot{\sigma} + \frac{1}{\mu}\sigma.\qquad(4.4.17)$$

Equation (4.4.17) represents the stress-strain response of a Maxwell model. On rearrangement, we obtain the following modified form of Eqn. (4.4.17), which is also known as the governing differential equation of a Maxwell model:

$$\sigma + \frac{\mu}{E}\dot{\sigma} = \mu\dot{e}$$

or

$$\sigma + \tau\dot{\sigma} = \mu\dot{e}\qquad(4.4.18)$$

where $\tau = \mu/E$ is called as the *relaxation time* (it has units of time). The significance of τ is discussed later in Example 4.4.1 below. Eqn. (4.4.18) can be integrated for different loading conditions, in order to simulate the various loading conditions such as creep, stress relaxation etc, that have been introduced in the beginning of Section 4.4 above.

Stress Relaxation and Creep Response of Maxwell Model

We will demonstrate the integration of the governing equation for Maxwell model based on stress relaxation loading condition in Example 4.4.1

Example 4.4.1: Evaluate the stress relaxation behaviour according to the Maxwell model. What is the significance of τ ?

Solution:

In stress relaxation, we know that a constant strain e^0 is applied and the stress response is observed. Substitution this constant value of strain in Eqn. (4.4.18), we get

$$\sigma + \tau \dot{\sigma} = 0 \qquad\qquad (E4.4.1)$$

Integrating Eqn. (E4.4.1), for the initial conditions of $\sigma = \sigma^0$ at $t = 0$, we get

$$\sigma = \sigma^0 \left(\exp\left\{ -\frac{t}{\tau} \right\} \right) = 0 \qquad\qquad (E4.4.2)$$

Based on Eqn. (E4.4.2), we can obtain the relaxation modulus according to Maxwell model as,

$$E_r(t) = \frac{\sigma(t)}{e^0} = \frac{\sigma^0}{e^0} \left(\exp\left\{ -\frac{t}{\tau} \right\} \right) \qquad\qquad (E4.4.3)$$

From Eqn. (E4.4.3), we can deduce that the relaxation time (τ) signifies the rate of exponential decay of the relaxation modulus. If τ is large, relaxation is slow and if τ is small, relaxation is relatively fast. Extreme values of τ, *i.e.*, $\tau = 0$ and $\tau = \infty$, imply dominant viscous and dominant elastic behaviour, respectively. Therefore, the relaxation time of the Maxwell model can be used to examine the relative importance of viscous or elastic response of a material. If $\tau = \infty$, we have

$$E_r(t) = \frac{\sigma^0}{e^0} = E \qquad\qquad (E4.4.4)$$

where we note that E is the instantaneous elastic modulus of the material.

Based on the discussion following Eqn. (4.4.18) and in Example 4.4.1, it is clear that the response according to the Maxwell model can reduce to elastic or viscous based on the value of its parameters. In Example 4.4.1, the relaxation time, instead of μ and E, was used to describe the limiting cases. It is to be noted that the physical significance of Maxwell model parameters E and μ should not be equated with the modulus and viscosity of the linear elastic and linear viscous materials, respectively. At most, these parameters can be used as indicators to analyze the relative importance of elastic like or viscous like behaviour according to the Maxwell model.

From Example 4.4.1, we note that the relaxation modulus for a Maxwell model with parameters E and τ, is given by

$$E_r(t) = E\left(\exp\left\{-\frac{t}{\tau}\right\}\right). \tag{4.4.19}$$

The stress response of a viscoelastic material, when subjected to varying strains (as opposed to the constant strain condition considered above), can be obtained by formulating the Maxwell model in an integral form. The integral form can be visualized as a linear superposition of the response associated with different strains, each of a different magnitude (discussed later in Chapter 5). Noting that Eqn. (4.4.18) is an ordinary differential equation, it can be written in an integral form (for example, using integrating factor approach):

$$\sigma(t) = e^{-\frac{t}{\tau}} \int_{-\infty}^{t} Ee^{\frac{t'}{\tau}}\dot{e}(t')dt' \text{ or } \sigma(t) = \int_{-\infty}^{t} Ee^{-\left(\frac{(t-t')}{\tau}\right)}\dot{e}(t')dt'. \tag{4.4.20}$$

In Eqn. (4.4.20), the overall time of the application of strain is considered. Hence, we use t' to monitor the time of application of strain from past ($t' = -\infty$) to present ($t' = t$). We can understand the integral in Eqn. (4.4.20) as stress responses to incremental strains. Magnitude of incremental strain $de(t')$ acting at any time, can be obtained as

$$de(t') = \dot{e}(t')dt'. \tag{4.4.21}$$

Therefore, the incremental stress produced by the incremental strain given in Eqn. (4.4.21), will be given by

$$d\sigma(t') = Ee^{-\left(\frac{(t-t')}{\tau}\right)}de(t'). \tag{4.4.22(a)}$$

The total stress response due to a strain history to the current time, assuming that the strain existed from $-\infty$ to the current time t, is then given by

$$\sigma(t) = \int_{-\infty}^{t} d\sigma(t') = \int_{-\infty}^{t} Ee^{-\left(\frac{(t-t')}{\tau}\right)}\dot{e}(t')dt'. \tag{4.4.22(b)}$$

Equation (4.4.22(b)) is called the *integral statement of the Maxwell model*. It also indicates the *fading memory* of viscoelastic response. The relaxation modulus is decreasing function of $(t - t')$. At times near the present time, the relaxation modulus is large. However, further back in past (*i.e.* larger $(t - t')$), the relaxation modulus is smaller. The incremental strain is multiplied by smaller value of modulus as $(t - t')$ increases. Therefore, the immediate past of the material has a greater contribution to $\sigma(t)$ compared to distant past. Hence, the material is assumed to have a fading memory. We note that general form of Eqn. (4.4.22) can be rewritten in terms of the relaxation modulus as

$$\sigma(t) = \int_{-\infty}^{t} d\sigma(t') = \int_{-\infty}^{t} E_r(t-t')\dot{e}(t')dt' . \tag{4.4.23}$$

Based on different models, relaxation modulus in Eqn. (4.4.23) will be different. Therefore, Eqn. (4.4.23) is referred to as general linear viscoelastic model.

If we subject Maxwell model to a creep test, we would apply a constant stress (say σ_0) to the model and would be monitoring the change of strain with time. Based on Eqn. (4.4.18), we can obtain $\dot{e} = \sigma_0 / \mu$. We note that this strain rate is the same as the strain rate that is predicted by a viscous model. However, the linear increase in strain that is predicted by this model is not in conformity with the exponential increase in strains that are observed in most creep experiments with viscoelastic materials. Hence, it is concluded that the Maxwell model is not appropriate to describe creep behavior in viscoelastic materials.

Oscillatory (Dynamic) Response of Maxwell Model

As was discussed in Section 4.4.1, the oscillatory response of viscoelastic materials is very important in describing their behaviour. The material functions such storage modulus, loss modulus, tan δ etc. were stated to be functions of frequency. In addition, they would depend on the model parameters. In the following discussion, we show how this dependence for Maxwell model.

Example 4.4.2: Derive the expressions for $\mu'(\omega)$ and $\mu''(\omega)$ based on the Maxwell model.

Solution:

In case of an oscillatory test, the strain rate applied is given by (for example, Eqn. (4.4.5)),

$$\dot{e} = \dot{e}^0 \cos \omega t . \tag{E4.4.5}$$

Substituting this in the equation for the Maxwell model (Eqn. (4.4.22)), we obtain

$$\sigma = \int_{-\infty}^{t}\left[\frac{\mu}{\tau}\exp\left\{-\frac{(t-t')}{\tau}\right\}\right]\dot{e}^{0}\cos\omega t'\,dt' \tag{E4.4.6}$$

Integrating Eqn. (E4.4.6) by parts, we get

$$\frac{\sigma}{\dot{e}^{0}} = \frac{\mu}{\tau}\left[\exp\left\{-\frac{(t-t')}{\tau}\right\}\int\cos\omega t'\,dt'\right]_{-\infty}^{t} - \int_{-\infty}^{t}\left[\int\left(\frac{\mu}{\tau}\cos\omega t'\right)dt'\right]\left\{\frac{1}{\tau}\exp\left\{-\frac{(t-t')}{\tau}\right\}\right\}dt'. \tag{E4.4.7}$$

Solving the integral and rearranging terms in Eqn. (E4.4.2.3), we obtain

$$\frac{\sigma}{\dot{e}^{0}} = \frac{\mu}{\tau}\left[\exp\left\{-\frac{(t-t')}{\tau}\right\}\frac{\sin\omega t'}{\omega}\right]_{-\infty}^{t} - \frac{\mu}{\omega\tau^{2}}\int_{-\infty}^{t}\sin\omega t'\exp\left\{-\frac{(t-t')}{\tau}\right\}dt'. \tag{E4.4.8}$$

Simplifying the first term of Eqn. (E4.4.8) and integrating by parts the second term, we get

$$\frac{\sigma}{\dot{e}^{0}} = \frac{\mu}{\tau}\frac{\sin\omega t}{\omega} - \frac{\mu}{\omega\tau^{2}}\left[\exp\left\{-\frac{(t-t')}{\tau}\right\}\frac{-\cos\omega t'}{\omega}\right]_{-\infty}^{t} - \frac{1}{\omega^{2}\tau^{2}}\frac{\mu}{\tau}\int_{-\infty}^{t}\cos\omega t'\exp\left\{-\frac{(t-t')}{\tau}\right\}dt'. \tag{E4.4.9}$$

Simplifying the second term of Eqn. (E4.4.9) and recognising that the integral in the third term is the same as in Eqn. (E4.4.6), we obtain

$$\frac{\sigma}{\dot{e}^{0}} = \frac{\mu}{\tau}\frac{\sin\omega t}{\omega} + \frac{\mu}{\omega^{2}\tau^{2}}\cos\omega t - \frac{1}{\omega^{2}\tau^{2}}\frac{\sigma}{\dot{e}^{0}}. \tag{E4.4.10}$$

Rearrangement of terms in Eqn. (E4.4.10) leads us to the expression

$$\frac{\sigma}{\dot{e}^{0}}\left(1+\frac{1}{(\omega\tau)^{2}}\right) = \frac{\mu}{\tau}\frac{\sin\omega t}{\omega} + \frac{\mu}{(\omega\tau)^{2}}\cos\omega t. \tag{E4.4.11}$$

Equation (E4.4.11) yields an expression for stress as a function of time, which is of the form

$$\sigma(t) = \dot{e}^0 \left[\frac{\mu}{1+(\omega\tau)^2}\cos\omega t + \frac{\mu\omega\tau}{1+(\omega\tau)^2}\sin\omega t \right] \tag{E4.4.12}$$

We note that in Eqn. (E4.4.10), the stress has been written as two contributions based on the Maxwell model. The first term containing $\cos \omega t$ signifies the viscous contribution as it is in phase with the loading (in this case, strain rate $\dot{e} = \dot{e}^0\cos\omega t$). The second term signifies the elastic response as it is out of phase with the strain rate. Based on the definition of μ' and μ'' (Eqn. (4.4.11)), we get

$$\mu'(\omega) = \frac{\mu}{1+(\omega\tau)^2}; \ \mu''(\omega) = \frac{\mu\omega\tau}{1+(\omega\tau)^2}. \tag{E4.4.13}$$

Let us examine the variation of $\mu'(\omega)$ and $\mu''(\omega)$ as derived in Eqn. (E4.4.13) with frequency and relaxation time. We note that $\omega\tau$ appears as a product in these expressions. We also note that ω is a loading parameter and τ is a material parameter. Let us consider the limiting cases of $\omega\tau$, namely $\omega\tau \gg 1$ and $\omega\tau \ll 1$. Based on Eqn. (E4.4.13), we notice that when $\omega\tau \ll 1$

$$\mu' \to \mu \ \text{and} \ \mu''(\omega) \to \mu\omega\tau$$
$$\tag{4.4.24}$$

Similarly, when $\omega\tau \gg 1$, we have

$$\mu'(\omega) = \frac{\mu}{(\omega\tau)^2} \ \text{and} \ \mu''(\omega) = \frac{\mu}{\omega\tau}. \tag{4.4.25}$$

When either the material relaxation time is very small or the loading frequency is very low (implying $\omega\tau \ll 1$), we observe that the dominant viscous response will be as $\mu' = \mu$.

We will now recall the storage modulus $E'(\omega)$ and loss modulus $E''(\omega)$ as given in Eqn. (4.4.10) and discuss their significance. Based on the response of Maxwell model to an oscillatory strain $(e = e^0 \cos\omega t)$ it can be shown that (Exercise 4.4)

$$E'(\omega) = \frac{\mu\omega^2\tau}{1+(\omega\tau)^2} \ \text{and} \ E''(\omega) = \frac{\mu\omega}{1+(\omega\tau)^2}. \tag{4.4.26}$$

Similar to the limiting cases illustrated in Eqns. (4.4.24) and (4.4.25), we obtain the limiting cases for the storage and loss modulii. The mathematical expressions for these modulii are given by the following expressions.

When $\omega\tau \ll 1$, $\quad E' \to \mu\omega^2\tau \ \text{and} \ E''(\omega) \to \mu\omega$, $\tag{4.4.27}$

when $\omega\tau \gg 1$, $E'(\omega) = \dfrac{\mu}{\tau} = E$ and $E''(\omega) = \dfrac{\mu}{\omega\tau^2} = \dfrac{E}{\omega\tau}$.   (4.4.28)

The limiting cases shown in Eqns. (4.4.27) and (4.4.28) have a special significance. At large loading frequencies, the storage modulus is equal to the Maxwell model parameter, E. Also, the magnitude of E'' will be much smaller than E'. This indicates predominant elastic response of the material, since the loss modulus is insignificant compared to the storage modulus and the storage modulus is independent of the loading frequency. At low loading frequencies (Eqn. (4.4.27)), $E' \propto \omega^2$ and $E'' \propto \omega$. It was shown in Eqn. (4.4.24), that at low loading frequencies $\mu' = \mu$, indicating viscous response. Hence, it is normally understood that the frequency dependence of the storage modulus according to Eqn. (4.4.27) implies a predominant viscous response of the viscoelastic material.

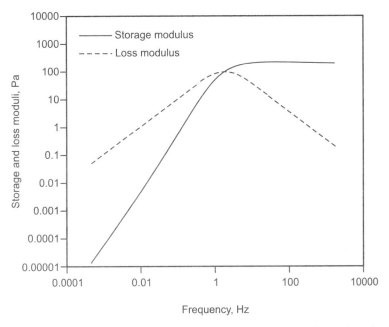

Fig. 4.5 Variation of storage and loss modulus with frequency for a viscoelastic material (μ = 100 Pa s, τ = 0.5 s^{-1})

The variation of E' and E'' with frequency can be used to demonstrate the viscoelastic behaviour described by the Maxwell model. As shown in Fig. 4.5, E' is larger at high frequency while E'' is larger at lower frequency. Therefore, at higher frequency elastic contribution dominates, while at lower frequency viscous behaviour dominates. The frequency at which the two moduli are equal is the inverse of the relaxation time. Near this frequency, $\omega\tau \sim 1$ and the material shows viscoelastic behaviour.

For materials with different relaxation times, the qualitative nature of the Maxwell response remains the same. As mentioned earlier, values of ωt can be used to observe the behaviour. Therefore, the limiting behaviour shifts to higher or lower frequencies, depending on the magnitude of the relaxation time. At a given loading frequency, a material with a larger relaxation time will show more dominant elastic behaviour compared to a material with a smaller relaxation time.

4.4.2.2 Kelvin-Voigt Model

As indicated in Section 4.4.2, in the Kelvin-Voigt, the kinetic responses of the elastic and viscous behaviour are added under the action of the same strain e. The stresses according to Eqns. (4.4.3) and (4.4.4) need to added, in order to get the total stress $\sigma(t)$. Hence, the governing differential equation for the Kelvin-Voigt model will be

$$\sigma = \mu\dot{e} + Ee . \tag{4.4.24}$$

The response of the Kelvin-Voigt to a creep experiment can be shown to be (Exercise 4.5(a))

$$e(t) = \frac{\sigma_0}{E}\left(1 - e^{-\frac{t}{\tau_{ret}}}\right), \tag{4.4.25}$$

where $\tau_{ret} = \mu/E$ is called the *retardation time*. Similar to relaxation time, the retardation time can also be used to examine the limiting behaviour of the Kelvin-Voigt model (Exercise 4.5(b)). It should be noted that the relaxation time and retardation time have the same expression (μ/E). However, the physical significance of the two times is different. Each of them is relevant only for the particular test condition and the model. It can be easily seen from Eqn. (4.4.24) that the response of Kelvin-Voigt model to stress relaxation is identical to an elastic response. Therefore, the Kelvin-Voigt model is not suitable for describing the stress relaxation of viscoelastic materials.

It will be helpful to examine the linearity (as defined in Section 4.1) of Kelvin-Voigt model for the specific case of creep loading. We recognize that the input variable is stress, while output variable is strain. We note that stress and strain are related to each other through a differential equation (Eqn. (4.4.24)). Addtionally, the strain response under creep loading is a function of time (Eqn. (4.4.25)). Hence, a scaled change in stress will result in a scaled change in strain (measured at given instant of time). In other words, strain resulting from a combination of stresses is the same as combination of strains (measured at the same instant of time) resulting from individual stresses (Exercise 4.6).

4.4.2.3 Mechanical Analogs for Viscoelastic Models

In literature, linear viscoelastic models are developed usually using mechanical analogs. A pictorial representation of the various mechanical analogs of viscoelastic models is shown in Table 4.1. In such a development, an elastic response is depicted by a spring (Table 4.1(a)) that behaves according to Eqn. (4.4.3). Similarly, a viscous response is depicted by a dashpot (Table 4.1(b)) that behaves according to Eqn. (4.4.4). Thus, the viscoelastic response is visualized as material behaviour resulting from a combination of spring elements and a dashpot elements.

The Maxwell model can be seen to be a combination of a spring and a dashpot elements *in series* (Table 4.1(c)), since the total strain response is the sum of strains of individual elements (with the stress being the same for both the elements). This mechanical analog of the Maxwell model, helps us to develop the governing equations (Eqn. (4.4.18)) based on the series combination of mechanical elements. It is easy to see that the response of the model, when a constant load is applied during a creep experiment, will be that of the dashpot alone, since the spring element will display only an instantaneous elastic response.

Kelvin-Voigt model is a combination of spring and dashpot elements *in parallel* (Table 4.1(d)), since the total stress is the sum of stresses in individual elements (with the strain being constant for both the elements). The governing equation for the Kelvin-Voigt model (Eqn. (4.4.24)) can be developed from parallel combination of mechanical elements.

Maxwell and Kelvin-Voigt models are limited because they do not capture multiple relaxation times exhibited by many viscoelastic materials. Additionally, the same viscoelastic material may be subjected to both creep and stress relaxation. It will be inappropriate to use different models for a single material for different loading conditions.Further, the instantaneous elastic response followed by overall viscoelastic response, observed in many materials, is not captured by Maxwell or Kelvin-Voigt model.Therefore, several other combinations of springs and dashpots have been used in the literature to simulate viscoelastic responses. For example, the *standard linear solid* or *Zener model* (Table 4.1(e)), also shown in Table 4.1, has a Maxwell model in parallel with another spring element. The simple viscoelastic model differential equation that captures the behaviour of this mechanical analog is given by (Exercise 4.5(a))

$$\sigma + \frac{\mu}{E_1}\frac{\partial \sigma}{\partial t} = E_2 e + \left(E_1 + E_2\right)\frac{\mu}{E_1}\frac{\partial e}{\partial t} \qquad (4.4.26)$$

The standard linear solid is one of the simplest models that captures the qualitative response of viscoelastic materials in creep, stress relaxation as well as oscillatory testing (Exercise 4.5(b)). Maxwell models in parallel (Table 4.1(f)) can be used to simulate the viscoelastic materials with multiple relaxation times (Exercise 4.6).

Table 4.1 Examples of Viscoelastic Models and their Mechanical Analogs

	Viscoelastic Model	Mechanical Analog	Main Applications
(a)	Hookean elastic solid		Metals, ceramics
(b)	Newtonian viscous fluid		Air, water, oils
(c)	Maxwell model		Stress relaxation, oscillatory testing (used for fluid-like viscoelastic materials)
(d)	Kelvin or Voigt mode		Creep (Used for solid-like viscoelastic materials
(e)	Jeffrey's model		Stress relaxation, oscillatory testing (used for fluid-like viscoelastic materials)
(f)	Standard linear solid (Zener model)		Stress relaxation, creep, oscillatory testing (used for solid-like viscoelastic materials)
(g)	Maxwell models in parallel		Stress relaxation, oscillatory testing (used for viscoelastic materials with multiple relaxation times)

4.4.3 Time Temperature Superposition

The effect of temperature on mechanical response of materials is very important for most applications. Most of the material parameters such as modulus and relaxation times are found to be functions of temperature. Since it is not always feasible to wait for long times at any given temperature to study the response of the material, accelerated testing is often used to assess the lifetimes of materials and components. Accelerated testing is usually done at higher temperatures than the temperature of normal use of the material. In a similar manner, for viscoelastic materials, *time temperature superposition* is used to relate mechanical response at different temperature. Time temperature superposition is possible for linear viscoelastic response at different temperatures based on the idea of time temperature equivalence. Since frequency (of loading) is inverse of time (larger frequency implies shorter times and smaller frequency implies larger times), there is also a frequency temperature equivalence of linear viscoelastic response.

Let us consider stress relaxation modulus for a Maxwell fluid (relaxation time τ_1) at a specified temperature θ_1. At another temperature θ_2, the stress relaxation modulus will be different because relaxation time (τ_2) of the material would be different. For current discussion, we will assume $\theta_1 > \theta_2$ and therefore $\tau_1 < \tau_2$. The relaxation modulus is depicted in Fig. 4.6. The relaxation modulus decreases as a function of time, as described in Eqn. (E4.4.3). Consider a particular value of relaxation modulus E_1, which is realized at time t_1 at temperature θ_1. The same value of relaxation modulus can be realized at temperature θ_2, provided we consider a different time. In other words, material property at a given temperature and time can be *found to be equivalent to* property at another temperature and also at a different instant of time. We can observe the relaxation modulus in a log-log plot (as shown in Fig. 4.6). It can be observed from the figure that the relaxation modulus at θ_1 can be *shifted to the left* to obtain relaxation modulus at θ_2. In other words, relaxation modulus measured at θ_1 can be used to obtain relaxation modulus at θ_2, if we modify the time scale suitably (implied in horizontal shifting).

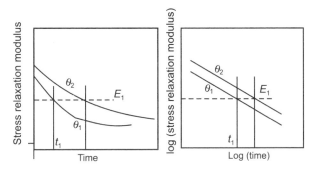

Fig. 4.6 Relaxation modulus based on Maxwell model at two different temperatures

In the above description, the discussion was restricted to Maxwell model. However, time temperature superposition is useful for linear viscoelastic materials in general, irrespective of which model is used to describe the mechanical response of the material. Measurements are normally carried out at different temperatures. Using shifting procedures, data is superimposed to obtain a *master curve* at a specified temperature.

SUMMARY

In this chapter, we recollected some of the simplistic models that are commonly used in solids and fluids. We noted that the simplistic models are also associated with a linear relation involving external loads and their responses. This linear response in solids is characterized by the Hooke's law of elasticity, which is a linear model between stresses and infinitesimal strains. We further noted that this model could be anisotropic in general and the popular isotropic version of the model has been documented. In fluids, a linear model is normally described between the stresses and the strain rates, through a Newtonian fluid model, which was recollected briefly in this chapter. The time dependent mechanical response of viscoelastic materials like polymers, was noted to be a function of stresses, strains and their rates. This response was also visualized to be a linear combination of a solid spring-like element and a flowing fluid-like element.

The contents of the topics discussed in this chapter will be expanded in the subsequent two chapters. In Chapter 5, we will discuss the non-linear viscous models that are being used in fluids. In Chapter 6, we will discuss the modeling of non-linear elastic solids both for infinitesimal as well as for finite deformations. Modelling of materials that respond to coupled fields (discussed in Chapter 7), will involve the use of simplistic models described in this chapter, in addition to additional terms and equations that exist because of the presence of non-mechanical fields (such as temperature, electric and magnetic fields).

EXERCISE

4.1 Consider a transformation between a coordinate system x_1, x_2, x_3 and x_1^*, x_2^* and x_3^*. The direction cosines between the various directions can be represented as given in the Table E4.1 below.

Table E4.1

	x_1	x_2	x_3
x_1^*	a_{11}	a_{12}	a_{13}
x_2^*	a_{21}	a_{22}	a_{23}
x_3^*	a_{31}	a_{32}	a_{33}

Consider the transformations $x_1^* = x_1$, $x_2^* = x_2$, $x_3^* = -x_3$, as was indicated in Section 4.2.2.2. Let $\Sigma_{\alpha\beta}$ represent the transformed stresses, which in the original co-ordinates are represented by $\sigma_{\alpha\beta}$. Similarly, let $E_{\alpha\beta}$ represent the transformed strains, which in the original co-ordinates are represented by $e_{\alpha\beta}$. It is possible to show that the transformation of stresses and strains between the two coordinate frames can be given by

$$\Sigma_{\gamma\delta} = \sigma_{\alpha\beta} a_{\gamma\alpha} a_{\delta\beta} \text{ and } E_{\gamma\delta} = e_{\alpha\beta} a_{\gamma\alpha} a_{\delta\beta} \quad \alpha, \beta, \delta, \gamma = 1, 2, 3$$

Perform the stress and strain transformations for the symmetric transformations indicated above and prove that for a general elastic body with one plane of elastic symmetry, there will need to be only 13 independent constants.

4.2 Consider an anisotropic material having the following properties:

$$E_x = 200\,\text{GPa}; \quad n = 2; \quad v_{xy} = 0.4; \quad v_{zx} = 0.3$$

(a) Assume that this material is being subjected to a uniform strain in the x direction. Evaluate all the stresses that are generated in the material. Compare your results with the behaviour of an equivalent isotropic material ($n = 1$ and $v_{xy} = v_{zx} = 0.3$).

(b) Assume that this material is being subjected to a uniform shear strain the xy plane. Evaluate the stresses that are generated in the material. Compare your results with the behaviour of an equivalent isotropic material.

4.3 Given the stress-strain relations given in terms of Lame's parameters in Eqn. (4.2.16) and the strain-stress relations given in terms of engineering constants in Eqn. (4.2.16). Prove the equivalence between the Lame's parameters and the Engineerng constants, which are given by

$$G = \mu, \quad E = \frac{G(3\lambda + 2G)}{\lambda + G} \quad \text{and} \quad v = \frac{\lambda}{2(\lambda + G)}.$$

(*Hint:* Subject the material to simple states of loading such as uniaxial tension, simple shear etc. to obtain the equivalence.)

4.4 (a) Show that the Maxwell model can be written for strain ($e(t)$) as follows,

$$\sigma = \int_{-\infty}^{t} \left[\frac{\mu}{\tau^2} \exp\left\{ -\frac{(t-t')}{\tau} \right\} \right] e(t,t')\,dt'$$

where $e(t, t')$ is the strain in the time interval between t' and t, given as

$$e(t,t') = \int_{t'}^{t} \dot{e}(t')\,dt'.$$

(*Hint:* Start with Eqn. (4.4.20) and integrate by parts.)

(*b*) Using the integral equation for the Maxwell model derived in part (a), show that the expressions for the storage and loss moduli are given by

$$E'(\omega)_\infty = \frac{\mu\omega^2\tau}{1+(\omega\tau)^2}; \; E''(\omega) = \frac{\mu\omega}{1+(\omega\tau)^2}$$

4.5 (*a*) Show that the creep response of the Kelvin-Voigt model is given by

$$e(t) = \frac{\sigma_0}{E}\left(1 - e^{-t/\tau_{ret}}\right).$$

(*b*) Explain the limiting behaviour of the above expressions for the following cases;

(*i*) $t \to 0$ and (*ii*) $t \to \infty$

4.6 Prove that the Kelvin Voigt model is a linear viscoelastic model, by examining its creep response, when subjected to scaled changes in the applied stress.

4.7 (*a*) Show that the differential equation associated with the response of a standard linear solid (shown in Table 4.1), is given by

$$\sigma + \frac{\mu}{E_1}\frac{\partial\sigma}{\partial t} = E_2 e + (E_1 + E_2)\frac{\mu}{E_1}\frac{\partial e}{\partial t}$$

(*Hint:* Recognize that the standard linear solid is a parallel combination of a spring and a Maxwell model. Hence apply the conditions for parallel combination of mechanical elements.)

(*b*) Show the response of the standard linear solid model can be given by,

(*i*) creep $e(t) = \dfrac{\sigma_0}{E_1} + \dfrac{\sigma_0}{E_2}\left(1 - e^{-t/\tau_g}\right)$; $\tau_g = \dfrac{\mu}{E_1 E_2/E_1 + E_2}$

(*ii*) stress relaxation $\sigma(t) = E_2 e_0 + (E_1 - E_2)e_0 e^{-t/\tau_g}$; $\tau_1 = \dfrac{\mu}{E_1}$

4.8 Derive the expression for the stress relaxation for a parallel combination of two Maxwell models as shown in Table 4.1(*f*).

■ ■ ■

5

Non-linear Models for Fluids

आपो वा इदगूं सर्व विश्वाभूतान्यापः प्राणा आपः पशव आपः ... (*महानारायणोपनिषत्*)
āpo vā idagṃ sarvaṃ viśvābhūtānyāpah prāṇā āpah paśava āpah ...
... (*Mahānārāyaṇ opaniṣad*)
This universe is indeed water, beings are water, vital airs are water,
animals are water...

5.1 INTRODUCTION

The general laws that govern the deformation of any material were outlined in
Chapter 3. We learnt that the governing laws are statements of mass balance,
linear momentum balance, angular momentum balance and energy balance.
As explained in Section 3.3.8, the equations that govern the above physical
principles can be solved only when they are used in combination with
constitutive relations. In Chapter 4, we examined some popular simplistic
material models for solids and fluids materials, which relate the stresses to the
kinematic quantities like strains or strain rates. These simplistic models were
linear in nature and are sufficient to model many of the engineering materials.
However, several materials do exhibit a non-linear behaviour in their mechanical
response and this behaviour needs to be modeled appropriately. In the current
chapter, we will examine the non-linear mechanical response of fluid materials,
while in Chapter 6, we will examine the non-linear mechanical response of
solid materials. The response of various engineering materials was described
as solidlike and fluidlike in Section 1.1.1. In current chapter as well as
subsequent chapters, we will use terms solid materials and fluid materials to
imply solidlike and fluidlike response, respectively.

We recall from Section 5.1 that linearity in constitutive relations refers to the
property that the output of a linear model from a combination of inputs is the
same as the combination of outputs from individual inputs. This property is

not observed for non-linear constitutive relations of engineering materials. The departure from linearity is discussed in detail with different types of material responses, both for fluids materials (in this chapter) and solids (in Chapter 6).

Non-linear models for capturing the non-linear behaviour, require higher number of material parameters in the mathematical description of the model. The estimation of additional parameters that are required to characterize these models through controlled experiments, remains a challenge for researchers and practitioners. These issues will be discussed with specific examples when we describe the case studies in current chapter and Chapter 6.

5.2 NON-LINEAR RESPONSE OF FLUIDS

In Chapter 4, we discussed the Newtonian fluids for which there is a linear relationship between stresses and strain rates. Any fluid for which such a relationship does not hold is generally treated as a *non-Newtonian fluid*. Broadly speaking, the mechanical behaviour of non Newtonian fluids not only depends on current levels of loads and deformations but also on the rates of loading and rates of deformation. In general, flow of fluids is also associated with large deformations.

A broad classification of non-Newtonian fluids can be as given below:

(*a*) *Non-linear viscous fluid*: A general non-linear viscous material is a material where the stresses are determined mainly based on non-linear relations of strain rates. Modeling of these materials is discussed in Section 5.3. Behaviour of lubricants during steady state operation is an example of such material behaviour.

(*b*) *Viscoelastic fluid*: Many other fluid materials also have an elastic response (for example, they can recover part of the deformation after removal of loading). These materials are called non-Newtonian viscoelastic materials. Polymer melt flow in complex geometries during plastics processing is an example where such behaviour is very important. We can classify the viscoelastic materials as follows:

(*i*) Linear viscoelastic material and

(*ii*) Non-linear viscoelastic material.

In Sections 4.1 and 4.4, linearity of the linear viscoelastic models was discussed in detail. As will be illustrated later in Sections 5.3 and 5.4, a scaled change in inputs will not result in scaled change in outputs in case of non-linear viscous as well as non-linear viscoelastic models.

5.2.1 Useful Definitions for Non-Newtonian Fluids

A set of material functions was defined in Section 4.4.1, as they are useful in describing material behaviour. However, additional material functions are defined to understand the complex behaviour of fluids. They can be devised either in shear mode or in extensional mode. As described in the Chapter 3, extensional and shear deformation imply nonzero diagonal and nonzero non-diagonal terms, (of deformation measures such as **D, E** and **e**) respectively. Similar to the discussion in Section 4.4.1, we will first describe the material functions in shear mode.

The simple shear flow was shown in Fig. 4.2. A material is placed between parallel plates, with one plate moving at a fixed velocity. Using measurements of stress, strain and strain rates under different conditions, we can define several material functions as described below. Some of these material functions are useful for describing both viscous as well as viscoelastic materials. Some others are useful for distinguishing between viscous and viscoelastic response.

5.2.1.1 Steady Shear

In steady shear, the material is subjected to a constant strain rate (D_{ij}) and the steady value of stress (τ_{ij}) is measured. The ratio of stress and strain rate is defined as the viscosity, which is given by the expression

$$\mu\,(D_{ij}) = \frac{\tau_{ij}}{2D_{ij}}. \qquad (5.2.1)$$

Thus defined, viscosity is called a material function, because the variation with strain rate may be described by different functional forms for different materials. Most of the materials (with strain rate dependent viscosity) exhibit strain rate independent viscosity at low shear rates. In other words, at low enough shear rates, most of the materials show Newtonian fluid behaviour. The strain rates, at which Newtonian behaviour is observed, vary from material to material. Equation (5.2.1) reduces to Eqn. (4.3.6) for Newtonian fluids. Some representative variations of viscosity with strain rate are shown in Fig.5.1. Steady viscosity, as defined in Eqn. (5.2.1) can be used for both viscous and viscoelastic fluids. However, unsteady nature of viscosity is useful in characterizing the viscoelastic fluids.

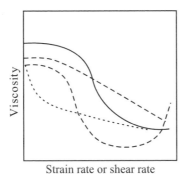

Strain rate or shear rate

Fig. 5.1 Some representative variations of steady viscosity with strain rate

Based on Eqns. (4.3.6), (4.4.4) and (5.2.1), we can observe that if a viscous material is subjected to a constant strain rate, the stress would also be constant. However, this is not the case with many other materials. In such materials, stress varies as a function of time when they are subjected to a constant strain rate. Only at the steady state, a constant value of stress is reached and we can use Eqn. (5.2.1) to define the steady viscosity. Therefore, a material can be subjected to constant strain rate and the stress can be monitored as a function of time. In such a case, time dependent viscosity can be defined as

$$\mu(t, D_{ij}) = \frac{\tau_{ij}(t)}{2D_{ij}} \tag{5.2.2}$$

The time dependent viscosity shown in Eqn. (5.2.2), is observed using two most common modes of experimentation, start-up and cessation. In start-up, the material is subjected to a constant strain rate at the initial time and then stress is monitored. In case of cessation, material is subjected to a constant strain rate till steady value of stress is reached. Subsequently, the deformation is stopped (strain rate equals zero) and the stress is monitored as a function of time. An example of response of materials for start-up is shown in Fig. 5.2. Stress increases and reaches a steady value in case of start-up. Therefore, this mode of experimentation is also called *stress growth* experiments. Similarly, cessation experiments are referred to as stress decay experiments. Such characterization is useful for simulating fluid behaviour during operational conditions such as start-up, switch off and transitions from one steady state to another during engineering applications.

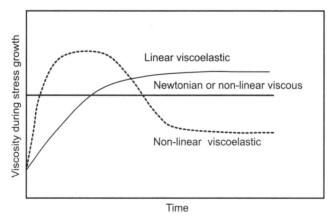

Fig. 5.2 Some examples of viscosity variation with time

Example 5.1: Evaluate the stress growth and stress decay response according to the Maxwell model. For a stress growth observation, consider the initial stress to be zero. For a stress decay observation, consider initial stress to be a constant.

Solution:

Maxwell model (see Section 4.4.2.1) is a linear viscoelastic model and is valid for small strains. Additionally, considering one dimensional flow, we can write

$$D_{ij} = \dot{e}_0. \tag{E5.2.1}$$

The governing differential equation for the Maxwell model is (see Eqn.(4.4.18))

$$\sigma + \tau\dot{\sigma} = \mu\dot{e} \tag{E5.2.2}$$

In a stress growth observation, the material is subjected to a constant strain rate. Therefore, we seek solution for the stress from the following equation

$$\sigma + \tau\dot{\sigma} = \mu\dot{e}_0. \tag{E5.2.3}$$

The solution for Eqn. (E5.2.3) can be written as

$$\sigma(t) = E\dot{e}_0 \int_0^t e^{-\left(\frac{t-t'}{\tau}\right)} dt',$$

or

$$\sigma(t) = \mu\dot{e}_0\left(1 - e^{-t/\tau}\right) \tag{E5.2.4}$$

Similarly, for a stress decay observation, we have

$$\sigma = \sigma_0\left(\exp\left\{-\frac{t}{\tau}\right\}\right); \sigma_0 = \mu\dot{e}_0, \tag{E5.2.5}$$

where $\dot{e}_0$ is the strain rate to which material was subjected and after reaching steady state, the strain rate was brought down to zero as an initial condition for the stress decay observation. The material functions based on the Maxwell model are given as

Stress growth: $\mu(t,\dot{e}_0) = \dfrac{\sigma(t)}{\dot{e}_0} = \mu(1 - e^{-t/\tau})$

Stress decay: $\mu(t,\dot{e}_0) = \dfrac{\sigma(t)}{\dot{e}_0} = \mu e^{-t/\tau}$ (E5.2.6)

Therefore, viscosity material function are exponentially increasing and decreasing for stress growth and stress decay respectively. The relaxation time, which is a very important material chacteristic, determines the rate of growth and decay. Therefore, time dependent viscosity can be used for examining the viscoelastic nature of fluids. It should be noted that viscosity, according to Maxwell model, though a function of time is independent of the strain rate.

5.2.1.2 Normal Stresses

Given the simple shear flow, the velocity profile can be evaluated to be

$$\mathbf{v} = D_{xy} y \mathbf{e}_x. \tag{5.2.3}$$

Stress for a Newtonian fluid can be evaluated as

$$\boldsymbol{\sigma} = \begin{bmatrix} -p & \tau_{xy} & 0 \\ \tau_{xy} & -p & 0 \\ 0 & 0 & -p \end{bmatrix} \tag{5.2.4}$$

It should be noted that the normal stresses τ_{xx}, τ_{yy} and τ_{zz} are zero. However, based on the requirements of material objectivity, the general expression for stress can be shown to be

$$\boldsymbol{\sigma} = \begin{bmatrix} -p+\tau_{xx} & \tau_{xy} & 0 \\ \tau_{xy} & -p+\tau_{yy} & 0 \\ 0 & 0 & -p+\tau_{zz} \end{bmatrix}. \tag{5.2.5}$$

Therefore, normal stresses may be present for simple shear flow of non-Newtonian fluids. Based on these normal stresses, two material functions are defined, *first normal stress difference* (N_1) *and second normal stress difference* (N_2):

$$N_1(D_{ij}) = \tau_{ii} - \tau_{jj}, N_2(D_{ij}) = \tau_{ii} - \tau_{kk}. \tag{5.2.6}$$

Clearly, the normal stress differences for simple shear flow of Newtonian fluid are zero. If N_1 and N_2 are non-zero, it is taken as one of the evidences for the elastic nature of fluids. Normal stress differences defined in Eqn. (5.2.6) are functions of strain rate. Similar to viscosity defined in Eqn. (5.2.3), normal stress differences can also be defined as functions of time and strain rate for start-up and cessation flows.

For defining all the material functions thus far, we have used shear flow as the controlled flow to which material is subjected. However, it is also possible to subject fluids to extensional flow. By definition, in extensional flows the diagonal components of velocity gradient and stretching tensors will be non-zero. This is analogous to tensile tests on solids, where the diagonal components of strain tensors are non-zero. Furthermore, it is possible to define uniaxial stretching, biaxial stretching etc. as the types of extensional flows based on the stretching tensor.

5.2.1.3 Material Functions in Extensional Flow

Extensional deformation is usually the default mode in which preliminary mechanical characterization of solids is carried out. Over the last decade, it has also become an extremely important mode for examination of material behaviour for fluidlike materials. These developments have been spurred due to inadequacy of analysis based on shear mode material functions. Satisfactory solutions for many design problems in engineering applications cannot be accomplished with shear mode material functions. For example, polymer fiber or film making operations (fiber spinning, film blowing etc.) involve large extensional deformations. In these operations, two materials with the similar shear mode material functions may lead to completely different behaviour. Similar to their shear counterparts (given in parenthesis in the following list), following are the most popular material functions used in extensional mode:

- Steady extensional viscosity or Trouton viscosity (Eqn. (5.2.1))
- Stress growth and stress decay viscosity (Eqn. (5.2.2))
- Creep compliance (Eqn. (4.4.1))
- Stress relaxation modulus (Eqn. (4.4.2))
- Complex modulus: storage and loss modulus (Eqn. (4.4.12))

A complete discussion on these extensional material function lies outside the scope of this book. However, we would like to highlight their importance in describing response of engineering materials.

5.2.2 Classification of Different Models

Several terms and classification schemes are used to describe various models that are used to describe the fluidlike materials. As described in Section 3.3, Newtonian fluid or linear viscous fluid is one of the most common models. If a material system exhibits response which cannot be described using Eqn. (3.3.5) for incompressible Newtonian fluids, we can term the material system as *non-Newtonian*. These *non-Newtonian* fluidlike materials or non-Newtonian fluids are classified further based on their response to applied loads. The response is most often discussed in terms of material functions described in Sections 4.4.1 and 4.2.

It should be noted that a single material itself can be considered as a Newtonian fluid or as different types of non-Newtonian fluids depending on the engineering application where the material is used. For example, we observed in Fig. 5.1, that depending on the strain rates, viscosity is constant or a function of strain rate. Therefore, depending on the strain rates of concern in the engineering application, the same material can be considered to have a constant viscosity (Newtonian fluid) or varying viscosity (non-Newtonian fluid).

It is possible that for an engineering application, viscosity of a material and its variation with strain rate are sufficient to have a mathematical description of the material response. This type of fluid behaviour can be completely characterized based on viscosity as the only material function. Materials and applications where this is possible, the material behaviour is termed as viscous. This behaviour and the models used to describe it are outlined in Section 5.3.

As mentioned in Section 3.4, viscoelastic materials response is based on the relations between stress, stress rates, strain and strain rates. It was also mentioned that at low strains the behaviour is termed as linear viscoelastic. We described one-dimensional simplistic models that are used to describe the linear viscoelastic response. In Section 5.4, we will describe the three dimensional constitutive relations for viscoelastic materials that are more general when compared with the models that were discussed in Section 4.4.

Materials are subjected to large deformation in various engineering applications. In fact, flow of all fluids involves very large deformations. In addition, materials are subjected to shear as well as extensional deformations in any typical engineering application. The non-linear viscous and linear viscoelastic models are inadequate to describe such mechanical response of materials. Various non-linear viscoelastic models have been proposed to describe the behaviour of materials undergoing large deformation. These models have been proposed either as extensions of linear viscoelastic models or based on requirements of objectivity that are demanded from material response functions in continuum mechanics (similar to those made in Section 3.4.4). The constitutive relations based on these models are sometimes written as differential equations, integral equations or their combination. Therefore, they are usually classified in terms of differential models or integral models and these models are described in Section 5.4.

5.3 NON-LINEAR VISCOUS FLUID MODELS

As mentioned earlier, constitutive relations for incompressible fluids will be described in this section as well as in the following sections. The generalised Newtonian fluids or non-Newtonian viscous fluids or purely viscous fluids are described based on the following constitutive relation (similar to Eqn. (4.3.5)):

$$\sigma_{ij} = -p\delta_{ij} + 2\mu(II_{\mathbf{D}})\, D_{ij}. \qquad (5.3.1)$$

Similar to Eqn. (4.3.6), we can write

$$\tau_{ij} = 2\mu\,(II_{\mathbf{D}})\, D_{ij}. \qquad (5.3.2)$$

Comparing Eqn. (5.3.2) with Eqn. (3.3.6), we can observe that the coefficient μ depends on the invariant of the stretching tensor ($II_{\mathbf{D}}$). As discussed in Section 4.4.2, material has characteristic response time (or several response times). Inverse of strain rate can be thought of experimental time scale. Based on differences in these time scales, different material response is obtained. Therefore, when material is subjected to different shear rates, we observe it to exhibit different viscosity. The second invariant of stretching tensor is a measure of strain rate. However, it should be noted that overall dependence of components is similar in linear viscous as well as non-linear viscous fluids. If a particular component D_{ij} is zero, then irrespective of the magnitude of μ, the stress σ_{ij} will also be zero. For example, in simple shear flow, all the normal stresses will be zero since all the diagonal components of **D** are zero. Another important feature of Eqn. (5.3.2) is that the stress depends on the instantaneous value of the stretching tensor. Therefore, no effect of history (or viscoelastic behaviour) can be described using this model. Due to these similarities with the Newtonian fluid, we also refer to this class of fluids as *generalized Newtonian fluids*.

The non-linear viscous fluid models can only be used if materials are being subjected to predominantly shear deformations. This usually implies material deformations with little or no change in geometry in a flow situation. Additionally, if the materials are being subjected to steady (deformation independent of time) flow, then these models can be used. These models are usually used in engineering design, when estimate of forces, pump power etc. are required. In many applications, a value for viscosity is required for this estimation. In these cases, the viscosity based on the non-linear viscous models can be used. These models, though useful in leading to good engineering estimates, are rarely useful for describing the details of material deformation such as velocity profiles or variation of stress or strain tensors with position.

Since these models are used largely for shear deformations, let us consider the simplification of Eqn. (5.3.2) for simple shear flow (as shown in Fig. 4.2). The constitutive relation for such a flow is given by

$$\tau_{xy} = 2\mu\,(D_{xy})\,D_{xy}. \tag{5.3.3}$$

Equation (5.3.2) is normally written in terms of a shear rate ($\dot{\gamma}$) defined as follows:

$$\dot{\gamma} = \frac{\partial v_x}{\partial y} \tag{5.3.4}$$

Based on Eqn. (5.3.4), we can identify that

$$\dot{\gamma} = 2D_{xy} \tag{5.3.5}$$

Since Eqn. (5.3.3) describes one dimensional flow, it is usually written in the following form:

$$\tau = \mu(\dot{\gamma})\dot{\gamma} \tag{5.3.6}$$

where τ_{xy} has been replaced by τ. In the following discussion, we will continue to use Eqn. (5.3.3) as the basis. Various empirical formulae are used for the description of viscosity as a function of strain rate. The choice of a model depends on the observed behaviour of the material in steady shear. Fig. 5.1 shows qualitative variation of viscosity with strain rate. If viscosity is found to be decreasing with increasing strain rate, the behaviour is termed as *shear thinning or pseudoplastic* behaviour. On the other hand, if the viscosity is found to be increasing with increasing strain rate, the behaviour is called shear thickening or *rheopectic* behaviour.

5.3.1 Power Law Model

One of the most commonly used expressions for viscosity is called the *power law* model for viscosity and is given by the expression

$$\mu(D_{ij}) = K(2D_{ij})^{n-1}, \tag{5.3.7}$$

where K and n are model parameters. This model is also referred to as Ostwald de waele model. The stress tensor according to the power law fluid model can be written as

$$\tau_{ij} = 2K(2D_{ij})^{n-1}D_{ij}. \tag{5.3.8}$$

More generally, Eqns. (5.3.7) and (5.3.8) can be written as

$$\mu\,(II_{\mathbf{D}}) = K(2II_{\mathbf{D}})^{\frac{n-1}{2}} \text{ and } \tau_{ij} = 2K(2II_{\mathbf{D}})^{\frac{n-1}{2}}D_{ij}. \tag{5.3.9}$$

5.3.2 Cross Model

Another example of non-linear viscous fluid model is the *Cross model* given as

$$\frac{(D_{xy}) - \infty}{0 - \infty} = \left[1 + (2t_r D_{xy})^{\frac{n-1}{2}}\right]^2. \tag{5.3.10}$$

where μ_0, μ_∞, t_r and n are model parameters. As shown in Fig. 5.1, viscosity of some fluids (such as polymer melts) is constant at very high and very low strain rates. Cross model can be used effectively to model both the high strain rate and low strain rate Newtonian plateaus. Power law, on the other hand, is suitable only at strain rates where viscosity decreases with strain rate.

As mentioned earlier in Section 5.2.2, these models can describe only the steady shear material function, viscosity and therefore are useful in applications where material deformation is predominantly steady shear. The response of these models to creep, stress relaxation and oscillatory testing is similar to Newtonian fluid response. Additionally, these models predict zero normal stress differences in steady shear. Though these models are usually introduced as important non-Newtonian models, they are less useful whenever details of deformation or stress fields are needed.

The linear viscoelastic models described in Section 4.4 are valid for small deformations and can describe the elastic behaviour of fluids. As mention in Sections 4.1 and 4.4, linearity in these models implied scaled output response to a scaled input response. The models described in this section, the non-linear viscous models, are useful to describe largely steady and mostly shear flows. They are non-linear because the output response is not scaled to a scaled input response.

The non-linear viscoelastic models, which are described in Section 5.4, can describe non-linear viscous response such as shear thinning as well as viscoelastic response at finite deformations.

5.4 NON-LINEAR VISCOELASTIC MODELS

Before discussing the non-linear viscoelastic models of fluids, let us briefly summarize the limitations of the linear viscoelastic models described in Section 4.4. In Section 4.4, we discussed the modelling of linear viscoelastic materials. Linear viscoelastic materials are very useful in describing small strain material behaviour in terms of material functions for creep, stress relaxation and oscillatory testing. Therefore, it was emphasized that in many engineering applications, the linear viscoelastic material functions can be used in preliminary characterization of material behaviour and in engineering design.

It can be shown that the response of many linear viscoelastic models such as Maxwell model (Exercise 5.1) to steady shear at constant strain rate reduces to Newtonian behaviour (constant viscosity). Several other linear viscoelastic models lead to a time dependent viscosity in steady shear (Exercise 5.1). Therefore, linear viscoelastic models are inappropriate to describe the strain rate dependence of viscosity described in Section 5.3. It can also be shown that normal stress differences based on linear viscoelastic models are zero (Exercise 5.1). Therefore, engineering applications where normal stresses maybe important, linear viscoelastic models cannot be used.

In summary, the non-linear viscous models and the linear viscoelastic models are inadequate for various engineering applications. Therefore, non-linear viscoelastic models have been proposed and are being developed to provide a better description of material behaviour. In this section, we list some of these models. It should be noted that these sections are not exhaustive in terms of the existing models. An attempt is made to discuss models which are in some ways extensions of those discussed earlier and also models which are popularly used in engineering applications.

As mentioned earlier, the non-linear models are described in terms of integro-differential equations. The most commonly used non-linear models are expressed in terms of either differential type viscoelastic models or integral models. The models of both these are generally referred to as rate-type viscoelastic models.

5.4.1 Differential-Type Viscoelastic Models

It can be shown that the three dimensional statement of Maxwell model given in Eqn. (4.4.18) is not frame invariant for large strains. This is due to the use of infinitesimal strain tensor in the development of Maxwell model. It was discussed in Section 3.4.5, that partial time derivative and material derivative of stress are not frame invariant. However, frame invariant model can be written based on the ideas of Maxwell model, if we write the stress rate in terms of *convected derivatives* as defined in Section 3.2.7.2. Starting from Eqn. (4.4.18), and replacing stress rate with convected rate and using stretching tensor instead of infinitesimal strain tensor, we obtain

$$\tau_{ij} + \tau \overset{\triangledown}{\tau}_{ij} = 2\mu D_{ij}. \tag{5.4.1}$$

Equation (5.4.1) is referred to as *upper convected Maxwell* model or *contravariant Maxwell* model.

Based on the definition of convected derivative given in Eqn. (3.2.88), we can rewrite Eqn. (5.4.1) as

$$\tau_{ij} + \tau \left[\frac{\partial \tau_{ij}}{\partial t} + v_k \frac{\partial \tau_{ij}}{\partial x_k} - \frac{\partial v_i}{\partial x_k} \tau_{kj} - \tau_{ik} \frac{\partial v_k}{\partial x_j} \right] = 2\mu D_{ij} \tag{5.4.2}$$

In Section 4.4.2.1, we observed that Maxwell model can be used to describe viscoelastic behaviour with two material constants, *i.e.*, *relaxation time* and *viscosity*. The elasticity of the material can be simulated using differing values of relaxation times. Since convected Maxwell models also contain only these two parameters, they are used in examining the effect of viscoelasticity on flow behaviour. In various cases, qualitative differences of the material response when compared to the Newtonian fluids can be explained based on the use of Maxwell model. Since they are the simplest among the viscoelastic models, convected Maxwell model is also used extensively to establish numerical procedures in computational fluid dynamics and finite element simulations. Many other models can also be viewed as extensions of the convected Maxwell model. Additionally, simplistic molecular theories lead to convected Maxwell models.

Convected Maxwell models (like their linear viscoelastic counterparts) have the same disadvantage in terms of describing the steady viscosity of viscoelastic materials, since they exhibit constant viscosity under steady shear or extension. However, convected Maxwell model does show non-zero normal stress difference.

We can also write the *upper convected Jeffreys* model as extension of the linear model (Table 4.1). This extension would yield

$$\tau_{ij} + \tau_1 \overset{\triangledown}{\tau}_{ij} = 2\mu \left(D_{ij} + \tau_2 \overset{\triangledown}{D}_{ij} \right) \tag{5.4.3}$$

This model is also referred to as the *Oldroyd B* model. This model contains another relaxation time as an additional parameter, when compared to the Maxwell models.

Another extension of Maxwell models is obtained by allowing the relaxation time and viscosity to be functions of stretching tensor. For this extension, we can write

$$\tau_{ij} + \tau(II_D)\overset{\triangledown}{\tau}_{ij} = 2\mu(II_D)D_{ij} \qquad (5.4.4)$$

This model is called *White Metzner model*. The number of parameters of White Metzner model depends on the functional forms to describe relaxation time and viscosity. For example, if a power law is used to describe both, it implies 4 parameters. Some examples of materials for which White Metzner model has been used for various fluids such as lubricants, electrorheological fluids, dough and polymer melts.

Many microscopic theories have been used to arrive at constitutive models for polymers and dispersed phase systems. The discussion about these microscopic theories is beyond the scope of this book. However, we note that these theories can be used to obtain a continuum constitutive model in terms of stress, strain and other variables. The most successful microscopic theory for polymer melts has been by Doi and Edwards. The following differential-type model is one statement of Doi-Edwards theory for polymer melts:

$$\tau_{ij} + \tau \overset{\triangledown}{\tau}_{ij} + \frac{2\tau}{3G}D_{kl}\tau_{lk}\tau_{ij} - G\delta_{ij} = 2\mu\, D_{ij}, \qquad (5.4.5)$$

where τ, G and μ are the parameters of the model. Compared to the convected Maxwell model (Eqn. (5.4.1), Eqn. (5.4.5)) contains additional contributions due to non-linear stress term (the third term on the left hand side). Models based on Doi-Edwards theory have been used to describe polymeric fluid flow in various geometries.

5.4.2 Integral-Type Viscoelastic Models

In Section 5.4.1, we learnt that upper convected Maxwell model is arrived at by replacing the partial time derivative with a convected derivative in the Maxwell model. Similarly, non-linear integral version of linear viscoelastic Maxwell model (Exercise 4.4) can be written as

$$\tau(t) = \int_{-\infty}^{t} \left[\frac{\mu}{\tau^2} \exp\left\{ -t - t'\!\!\Big/_{\tau} \right\} \right] \mathbf{E}_t(t,t)dt' \qquad (5.4.6)$$

The non-linear counterpart of the general linear viscoelastic integral Eqn. (4.4.23) can be written as

$$\tau(t) = \int_{-\infty}^{t} M_r(t - t')\mathbf{E}_t(t')dt' \qquad (5.4.7)$$

The constitutive relation given in Eqn. (5.4.7) is called the *Lodge rubberlike liquid* model.

We have described several classes of models for fluids. In Section 5.5, we will outline the applications of various models to describe mechanical response of a specific material, asphalt or bitumen.

5.5 CASE STUDY: MECHANICAL BEHAVIOUR OF ASPHALT

Sections 5.1–5.4 have outlined various modelling strategies to describe the behaviour of fluids. The use of some of the strategies to model a specific material is discussed through a case study on asphalt. Initially, the material description and experimental procedures for characterization of asphalt are given. The detailed description of a model and comparison with observed response is presented finally.

5.5.1 Material Description

Asphalt is widely used for pavement and waterproof coating applications. It is the bottom fraction from peteroleum processing and is a multi-component and multi-phase mixture. It contains thousands of different types of molecules, differing in their chemical nature in terms of polarity, stability etc. The different molecules are distributed in relatively dispersed phases such as *asphaltenes, resins* and *crystallizable fractions*. Because of the nature of underlying constituents and their interactions with each other, asphalt behaviour is different depends on the temperature, loading conditions and environmental conditions.

It is very important to understand the relationship between the chemical constituents of asphalt, their microstructure and their influence on the deformation response of the material. In engineering applications, asphalt is usually used with other materials such as sand, gravel, polymers etc. However, it is important to understand asphalt behaviour so that behaviour of asphalt based engineering materials can be analyzed.

We note that asphalt will be subjected to temperature ranges of –20°C to 60 °C, depending on weather conditions. On the other hand, asphalt is subjected to 150–200°C during construction of pavements. Given the structural requirements of pavement applications, deformation behaviour of asphalt is very crucial in determining the pavement performance. Mechanical response of asphalt can be in any one of the following ways, depending on the external temperature to which it is exposed, as illustrated in Table 5.1.

Table 5.1 Mechanical Response of Asphalt at Different Temperatures

S. No.	Temperature Range	Mechanical Behaviour
1.	Below –10 °C	Linear elastic (small deformation) or plastic (large deformation) behaviour
2.	–10 °C – 100 °C	Viscoelastic behaviour
3.	Above 100 °C	Newtonian (or linear viscous) behaviour

As was mentioned earlier, constituents such as asphaltenes, resins, saturated and unsaturated hydrocarbons (among them crystallizable waxes) are crucial in determining the asphalt behaviour. It is difficult to propound the microstructure of asphalt with certainty because of its complex nature. Based on current understanding, all the constituents such as asphaltenes, resins, crystallizable fractions are assumed to form aggregates. Alternately, it is assumed that these constituents are interconnected with each other through network like structures. Response of these microstructures depends not only on stress and strain, but also on their rates. Temperature plays an important role in deciding the overall behaviour because of thermal energy of molecules. At very low temperatures, most molecules/constituents are "frozen". While at very high temperatures, most molecules/constituents have sufficient thermal energy to prevent formation of aggregates/network. Therefore, as described in Table 5.1, the asphalt behaviour at very low and very high temperatures can be described by simplistic models such as linear elastic and linear viscous models, respectively.

Models of linear elasticity and Newtonian fluids are discussed in Chapter 4 and they are used for asphalt in the temperature ranges indicated. Models of plasticity are discussed in Section 6.4, and they are also sometimes used for describing response of asphalts at large deformation. In the intermediate range of temperature, (−10 °C – 100 °C), asphalt behaviour has been described by models discussed in Sections 4.4, 5.3 and 5.4. In Sections 5.5.2 – 5.5.4, such asphalt models are briefly described.

5.5.2 Experimental Methods

Rheology is defined as the science of deformation and flow of matter. It is the study of the manner in which materials respond to applied stress or strain. Instruments which measure rheological properties are called rheometers. Common rheometers, capable of measuring viscoelastic characteristics of fluids may be classified into two general categories: rotational and tube (or capillary) type.

In rotational rheometers, the sample under study is sandwiched between two surfaces; one of these is fixed and the other rotated. Rotational rheometers can be operated in different modes such as steady shear (constant angular velocity), oscillatory (dynamic), creep and stress relaxation modes. There are three traditional type of fixtures that are used to measure the viscous / viscoelastic properties of materials. They are: (*a*) Parallel plate viscometers, (*b*) Cone type viscometers and (*c*) Plate and concentric cylinder type viscometer. These viscometers are schematically shown in Fig. 5.3. For solidlike materials, parallel plate or cone and plate fixtures are used and for dilute solutions the cylindrical geometry is used.

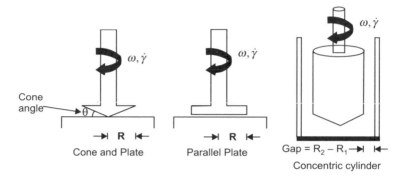

Fig. 5.3 Geometries used in a rheometer

In the temperature range of use of asphalt, asphalt shows significant solid-like behaviour and therefore, most common geometries of viscometers that may be used to investigate ashphalt viscoelasticity are parallel plate viscometers and cone and plate viscometers. It should be noted that sample preparation stage itself involves applied deformation to the material. Therefore, uniform conditioning should be used before measuring different material functions. Conditioning usually implies exposure to a preset temperature and/or deformation, before making the measurements to characterize material behaviour. In the case of a complex material such as asphalt, conditioning is very important in obtaining reliable results.

In a stress growth experiment, material is subjected to a constant strain rate (as shown in Fig. 5.2). The response of the material in the form of stress or stress growth viscosity is monitored as a function of time. The stress or stress growth viscosity evolves as a function of time eventually reaching a constant value, the steady stress or steady viscosity, respectively. At very low strain rates, the overall cumulative strain is small and therefore, linear viscoelastic response is observed as a limiting behaviour.

5.5.3 Constitutive Models for Asphalt

Several attempts have been made to capture the mechanical response of asphalt. As will be seen later, the material exhibits some basic features of a non-linear viscous deformation along with some indicators of elastic recovery. Hence, several attempts have been made to model the mechanical response in the past. These attempts can broadly be classified as (*a*) Non-linear viscous models, (*b*) Linear viscoelastic models and (*c*) Non-linear viscoelastic models. We will attempt to give a brief description of each of these models in this review below.

5.5.3.1 Non-linear Viscous Models

As was discussed in Sections 5.2.1.1 and 5.3, steady viscosity is used to characterize deformation behaviour of fluids. Viscosity of different asphalts is measured at temperatures around 60°C. As described in Table 5.1, at much lower temperatures (below −10°C), asphalt is very solidlike and therefore, viscosity cannot be measured. At higher temperatures (above 100°C) viscosity is constant (Newtonian fluid behaviour). The variation of viscosity with strain rate is captured using several non-linear viscous models. These models and their special features are summarized briefly in Table 5.2.

Table 5.2 Common Non-linear Viscous Models Used for Asphalt

S No	Model	Features
1	Power law or Ostwald de Waele model	Non-Newtonian viscosity, shear thinning behaviour, viscosity in extension[1]
2	Sisko model	Constant viscosity at low strain rates followed by shear thinning at higher strain rates
3	Cross model	Constant viscosity at low and high strain rates with shear thinning in intermediate strain rates[2]

The non-linear viscous models can be used for describing steady viscosity of asphalts. These are also used to indicate overall performance of asphalts. Given the varied nature of loading and temperature conditions, steady viscosity is a poor indicator of the overall mechanical behaviour of asphalt. Therefore, time dependent mechanical behaviour is examined using viscoelastic characterization. As described in Sections 5.5.3 and 5.5.4, linear as well non-linear viscoelastic models have been used to capture mechanical response of asphalt.

5.5.3.2 Linear Viscoelastic Models

Pavement is subjected to time dependent loading. Since asphalt exhibits viscoelastic response under most application conditions, it is very difficult to simulate complete mechanical response. In laboratory conditions, asphalt is characterized in oscillatory, creep or stress relaxation loadings. Linear viscoelastic models are used to describe these experimental characteristics.

---

[1] Cheung, CY and Cebon D: Experimental study of pure bitumens in tension, compression and shear, Jounral of Rheology, 41(1) (1997) 45-73.

[2] García-Morales M, Partal P, Navarro FJ, Martinez-Boza F, Mackley MR, Gallegos C: The rheology of recycled EVA/LDPE modified bitumen, Rheologica Acta, 43 (2004) 482–490.

The most common model that is used to describe the linear viscoelastic behaviour of asphalt, is the Burger's model. The mechanical analog of this model can be visualized as a series combination of Maxwell and Kelvin models (see Exercise 5.2 for mechanical analog). The governing equation for the model is given as follows:

$$\sigma + \left(\frac{\eta_1}{E_1} + \frac{\eta_1}{E_2} + \frac{\eta_2}{E_2}\right)\frac{\partial\sigma}{\partial t} + \frac{\eta_1\eta_2}{E_1E_2}\frac{\partial^2\sigma}{\partial t^2} = \eta_1\frac{\partial e}{\partial t} + \frac{\eta_1\eta_2}{E_2}\frac{\partial^2 e}{\partial t^2}, \qquad (5.5.1)$$

where η_1, η_2, E_1 and E_2 are parameters of Burger's model. This model can be visualized as a combination of Maxwell model and Voigt model (see figure accompanying Exercise 5.2). η_1 and E_1 correspond to the Maxwell model contribution and η_2 and E_2 correspond to the Voigt model contribution. Unlike most of the linear viscoelastic models discussed in Chapter 4, this model has second derivatives of stress and strain. It shows an elastic jump as soon as creep loading is applied (see Exercise 5.3). Such an elastic jump is observed with many asphalts. This is one among many features of asphalt behaviour, that are described by Burger's model.

In Sections 4.4.2.2 and 3.2.2.3, it was discussed that many engineering materials have multiple relaxation times. The set of all the relaxation times of a material and its distribution is called the relaxation time spectrum. It was mentioned that linear viscoelastic models incorporating multiple relaxation times can be considered to be parallel combination of Maxwell model (for example), with each Maxwell model representing one relaxation time. Several relaxation time spectrum models have been proposed for asphalt. Unlike the Burger's model discussed earlier, the spectrum can include a large number of relaxation times and therefore describe asphalt behaviour in a better way. Usually the relaxation time spectrum is obtained based on a master curve (described in Sections 4.4.2 and 4.4.3). Response of asphalt at different temperatures in the linear viscoelastic regime is measured. Then, using time temperature superposition, a master curve at a specified temperature is constructed. This master curve can be used to evaluate the relaxation time spectrum of the material[1]. However, these linear models, by definition, cannot capture the non-linear rheological behaviour of asphalt.

5.5.3.3 Non-linear Viscoelastic Models

In recent years, non-linear viscoelasticity of asphalt is being investigated. Normal stress differences, creep and stress relaxation are being observed at larger loading. Stress growth and cyclic experiments are being conducted to

---

[1] Lesueur D, Gerard J-F, Claudy P, Letoffe J-M, Planche J-P, Didier P: Structure related model to explain asphalt viscoelasticity, Journal of Rheology 40 (1996) 813-836

examine the non-linear viscoelastic behaviour of asphalt. To understand this behaviour, very few non-linear viscoelastic models have also been developed and these are summarized in Table 5.3.

Table 5.3 Examples of Non-linear Viscoelastic Models Used for Asphalt

S.No.	Model	Features
1	Constitutive model of asphalt based on evolving natural configurations[1]	Model was shown to explain creep behaviour of asphalt in the non-linear viscoelastic regime. For small deformations, one dimensional version of this model reduces to the Burger's model (Eqn. (5.5.1))
2	Lodge rubberlike liquid[2]	With chosen memory function (function of time and strain), viscosity and relaxation modulus of asphalt based materials were modelled.
3	Structural White Metzner Model[3]	Model uses a microstructural parameter. Model was shown to describe stress growth and cyclic behaviour. The model reduces to White Metzner model if structural parameter is kept constant

Stress growth (as described in Section 5.2.1.1) under steady shear can be investigated to examine the non-linear viscoelastic nature of asphalt. With repeated tests, asphalt has also been shown to exhibit different behaviour. Therefore, repeated tests on stress growth are very useful in determining the non-linear viscoelastic behaviour of the material. In the following discussion, the structural White Metzner model listed in Table 5.3. By assuming dependence of material behaviour on microstructure, the following extension of White Metzner model (Eqn. (5.4.4)) is proposed:

$$\tau + \tau(II_\mathbf{D}, \lambda)\overset{\triangledown}{\tau} = 2\,\mu(II_\mathbf{D}, \lambda)\mathbf{D} \qquad (5.5.2)$$

where λ is a structural parameter. In Material Description section of Section 5.5.1 described earlier, it was mentioned that asphalt is a multi-component multiphase mixture. These different components/phases give rise to a distinct microstructure of asphalt. This microstructure is crucial in determining the mechanical response of asphalt. Therefore, it is important to

---

[1] Krishnan JM, Rajagopal KR: On the mechanical behaviour of asphalt, Mechanics of Materials, 37 (2005) 1085–1100

[2] Polacco G, Stastna J, Biondi D, Zanzotto L: Relation between polymer architecture and non-linear viscoelastic behaviour of modified asphalts, Current Opinion in Colloid & Interface Science, 11 (2006) 230–245

[3] Vijay R, Deshpande AP, Varughese S: Nonlinear rheological modeling of asphalt using White-Metzner model with structural parameter variation based asphaltene structural build-up and breakage, Applied Rheology 18 (2008) 23214-23228

understand the microstructure and relate it to the mechanical response. By writing Eqn. (5.5.2), it is being assumed that viscoelastic behaviour of asphalt is dependent on microstructure and the state of microstructure can be captured using a parameter, called the structural parameter.

In this conception of microstructure, it is assumed that asphalt has a rest structure and this structure is modified when it is subjected to large deformations. Hence, λ changes depending on the deformation of asphalt. It is assumed that structure results from *formation* (of linkages of constituents) and gets modified due to *breakage* (of linkages of constituents). The influence of formation and breakage processes on λ is described through the following equation[1],

$$\frac{d\lambda}{dt} = a(1-\lambda)^b - c\lambda(II_D)d, \qquad (5.5.3)$$

where, the first term and second term on the right hand side are for formation and breakage of structure, respectively. The rate constants for the structural break-up and building are c and a, respectively. Using Eqns. (5.5.2) and (5.5.3), non-linear viscoelastic behaviour of asphalt can be described, provided that the functional forms for τ and μ in Eqn. (5.5.2) are specified.

As an example result from the model, the stress growth at two strain rates is shown in Fig. 5.4. At low shear rates (Fig. 5.4(a)), the behaviour of asphalt is very similar to linear viscoelastic material and therefore, stress growth is almost monotonic. The viscosity increases and reaches the steady state. On the other hand, viscosity at higher strain rate increases, decreases and then reaches a steady state (Fig. 5.4(b)). It should be noted that the steady viscosity is lower (of the order of 200 units) at higher strain rate, showing the shear thinning behaviour. A non-linear viscous model can describe this shear thinning nature, but cannot describe the stress growth shown in Fig. 5.4(a, b).

Qualitatively different stress growth behaviour at the two strain rates is observed due to structural changes taking place in the material. Based on solution of Eqn. (5.5.3), the structural parameter variation can be evaluated as a function of time. Due to shearing, structural changes lead to change in λ, as shown in Fig. 5.4(c). In the case of low strain rate, the structural parameter does not change significantly. Since low strain rate implies sufficient time for forming processes, even though breakage processes take place. Hence, the structural parameter does not change significantly during shearing. At large strain rates, significant breakage takes place, with less time for forming processes. Therefore, at steady state the structural parameter is very low. From Fig. 5.4(c), we can also observe that at higher shear rate, the structural parameter decreases rapidly, before reaching the steady state.

---

[1] Vijay R, Deshpande AP, Varughese S: Non-linear rheological modeling of asphalt using White-Metzner model with structural parameter variation based asphaltene structural build up and breakage, Applied Rheology 18 (2008) 23214-23228

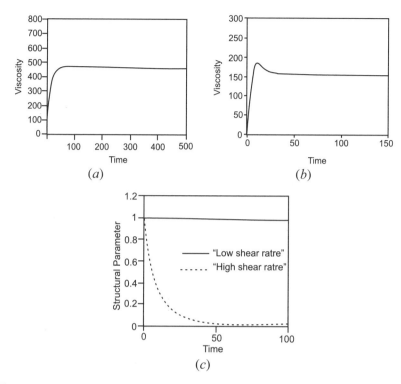

Fig. 5.4 Viscosity (Pa s) and structural parameter with time (s) at different shear rates (a) Viscosity at low shear rate (b) Viscosity at high shear rate (c) Structural parameter at low and high shear rates

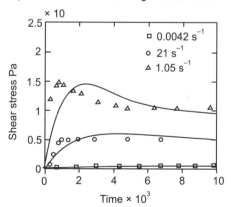

Fig. 5.5 Comparison of experimental data and model description: stress growth at different strain rates[1], (experimental data[2]).

[1]Vijay R, Deshpande AP, Varughese S: Non-linear rheological modelling of asphalt using White-Metzner model with structural parameter variation based asphaltene structural build-up and breakage, Applied Rheology 18 (2008) 23214-23228

[2] Vinogradov GV, Isayev AI, Zolotarev VA, Verebskaya EV: Rheological properties of road bitumens. Rheologica Acta 16 (1977) 266-281

As an example of the applicability of including structure based constitutive model, Fig. 5.5 shows comparison between model predictions and experimental data. As can be seen in the figure, at different shear rates the overshoot in stress as well as the steady value of stress is described well by the model.

Based on the history of deformation, asphalt steady shear behaviour has been observed to be different. The variation of viscosity with time for repeated testing can be used to monitor change in asphalt behaviour. The model given in Eqns. (5.5.2) and (5.5.3) can be used to understand the behaviour in terms of variation in microstructure. For example, variation of viscosity and structural variation can be monitored at a given shear rate after repeated testing. Between the repeated tests, the material is subjected to very high rates of shearing, to effect most changes in structure.

Results of such repeated tests according to the model are shown in Fig. 5.6. During the first test, the material starts from rest and therefore, the structural parameter is 1. With shearing, it decreases due to breakage of structure. At the same time, viscosity increases, decreases and reaches a steady state. If the material is sheared at very high shear rates for some time, even more structural breakdown takes place leading to lower value of structural parameter. After this shearing at very higher rate, second test on the material is carried out. We can observe that viscosity increases and reaches the steady state. Similarly, structural parameter also increases and reaches steady state. It should be noted that steady values of viscosity as well as structural parameters are same in the first and second test. However, their evolution with respect to time is entirely different. Second and third tests lead to almost the same results.

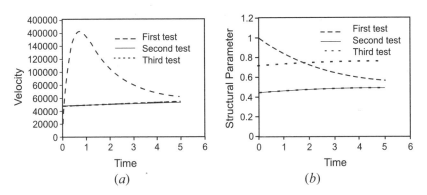

Fig 5.6 Variation of (a) viscosity and (b) structural parameter after repeated tests

Based on the results discussed with the model, we can observe that it is possible to describe non-linear viscoelastic behaviour of asphalt and related materials such as modified asphalts. The major advantage of the model is the possibility of relating the behaviour to asphalt composition through structural parameter. However, it is challenging to define metrics for composition and relate them to a structural parameter. This case study has highlighted various models which are used to model different aspects of mechanical response of asphalt. Additionally, a specific model was described in detail. Through the discussion of this model, we observed how an existing modelling strategy could be modified to describe detailed mechanical response of an engineering material such as asphalt.

SUMMARY

Engineering applications of various fluidlike materials involves description of their mechanical response. Linear viscous model or Newtonian fluid is useful for various fluids, but inadequate for large classes of fluids such as paints, pastes, polymeric fluids, asphalt, blood etc.

Several material functions were described that can be used to characterize the deformation behaviour of fluids. These include steady viscosity, stress growth and normal stress difference. Non-linear viscous models are very useful for describing the steady viscosity. Power law and Cross model are examples of such models and they can be used to describe the shear thinning of fluids. However, in time dependent deformations, non-linear viscoelastic behaviour of fluids is relevant. Differential and integral type non-linear viscoelastic models used for such cases were described briefly. Important examples of this class of models are convected Maxwell and White Metzner model.

Finally, a case study on deformation behaviour of asphalt was presented. Material description was followed by brief outline of experimental methods. Case study also included several models which have been found useful in describing non-linear viscous and linear and non-linear viscoleastic behaviour of asphalt.

EXERCISE

5.1 Evaluate steady shear viscosity and normal stress differences according to the following models:

(*a*) Maxwell model

(*b*) Jeffrey's model

(*c*) Upper convencted Maxwell model

(*d*) Upper convected Jeffrey's model

5.2 Figure 1 given below, shows mechanical analog of Burger's model. Show that the governing equation of Burger's model is given by (Eqn. (5.5.1))

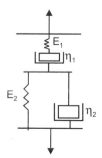

5.3 Figure (*a*) given below, shows creep loading and Figure (*b*) shows a representative response of asphalt to the loading. Start from the governing equation of Burger's model (Eqn. (5.5.1)) and show that Burger's model can show the given response.

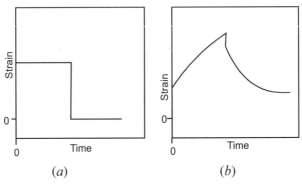

(*a*)   (*b*)

5.4 Consider combination of Maxwell model and standard linear solid model in series. Evaluate the creep response for the combination.

5.5 Another model which describes the creep behaviour in extension is similar to the Burger's model and is called the Dorn model. The Dorn model is an expression for the strain response of a fluid to creep loading:

$$e(t) = e_e + \dot{e}_{ss}\, t + e_T (1 - e^{-\Lambda \dot{e}_{ss} t}),$$

where e, $\dot{e}$, e_T and Λ are model parameters. e_e is the instantaneous elastic strain (time = 0), $\dot{e}_{ss}$ is the strain rate at steady state. The term involving $\dot{e}_{ss}$ implies viscous (Newtonian) behavior at large times (or at steady state). Show that the creep response given by the above equation is same as answer to Exercise 5.4.

■ ■ ■

6

Non-linear Models
for Solids

मृत्तिके देहि में पुष्टिम् त्वयि सर्वं प्रतिष्ठितम् ।   (महानारायणोपनिषत्)

mṛttike dehi me puṣṭiṃ tvayi sarvaṃ pratiṣṭhitam.
(Mahānārāyaṇopaniṣad)

O Earth, everything is established in you. Hence, give me nourishment.

6.1 INTRODUCTION

In Chapter 5, we described the non-linear mechanical response of fluidlike materials. In this chapter, we will discuss the non-linear mechanical response of solidlike materials. We recall from Section 4.2, non-linear solidlike material response can either be non-linear elastic response or non-linear inelastic response. Elastic response implies that materials retrace their path when unloaded. *Hyperelastic* and *Cauchy elastic* models that will be discussed in this Chapter, describe the non-linear elastic behaviour of materials. Similarly, we will also discuss *hypoelastic* models, which fall under non-linear inelastic models.

6.2 NON-LINEAR ELASTIC MATERIAL RESPONSE

Some of the common solids that exhibit non-linear mechanical response, are rubberlike materials. Examples of these materials are natural rubbers and elastomers (artificial rubbers) and biological gels. These materials exhibit finite deformations, with extension ratios as large as 800%.

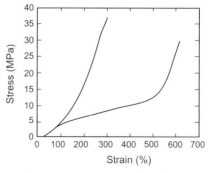

Fig. 6.1 Some examples of large elastic non-linear deformations

An example of large elastic non-linear deformations is shown in Fig. 6.1.

We have already seen in Chapter 4 that 'elasticity' is characterized by a property that the materials return to their original state of deformation when the external forces are removed, following the same path taken during loading. Mathematically speaking, this would also mean that there will exist a unique correspondence between a kinematic measure and the kinetic measure. We have seen in Section 3.4.4.8 (Eqn. (3.4.27)) that properties of material frame indifference would allow us to use the right Cauchy Green deformation tensor $\mathbf{C}$ to monitor the deformation of elastic materials. Similarly, first Piola-Kirchhoff stress $\mathbf{S}$ can be used to monitor the stresses in the material. Hence, an elastic material is characterized by a constitutive relation given below:

$$\mathbf{S} = \hat{f}(\mathbf{C}) \qquad (6.2.1)$$

Further, the unique correspondence between $\mathbf{S}$ and $\mathbf{C}$ requires that there exists a unique inverse given by

$$\mathbf{C} = \hat{g}(\mathbf{S}) \qquad (6.2.2)$$

Comparing with Eqn. (2.4.10), we observe that while writing Eqn. (6.2.1), we have assumed isothermal condition. Therefore, stress is not dependent on temperature and gradient of temperature.

We observe that Eqn. (6.2.1) is a serious restriction on the shape, that a stress-extension curve can take, for an elastic material. While Eqn. (6.2.1) allows for non-linear relation between $\mathbf{S}$ and $\mathbf{C}$, it allows for only monotonic change in the curve, preventing humps or valleys of any kind. We note additionally that in Eqns. (3.4.27) and (6.2.1), stress does not depend on the rates of deformation (for example, $\dot{\mathbf{C}}, \dot{\mathbf{U}}, \dot{\mathbf{F}}$). Therefore, it is implied that the current stress depends only on the present state of deformation and not on the deformation history. This would ensure that the retracing of the path will occur when material is subjected to cycles of loading and unloading.

Constitutive relations where functional forms given in Eqns. (6.2.1) and (6.2.2) are derived from a general statement of internal energy or strain energy density, are called as *hyperelastic* or *Green-elastic* material models. Constitutive relations for elastic materials which work on the explicit statements of the type given in Eqn. (6.2.1) are termed as *Cauchy elastic* material models. Both these models could be expressed using either finite deformation measure such as $\mathbf{C}$ or an infinitesimal deformation measure such as $\mathbf{e}$. Both these expressions involve the total deformational measures as opposed to incremental deformational measures, which will be introduced later as *hypoelastic models*

(Section 6.3). A detailed description of hyperelastic models and Cauchy models is given in Sections 6.2.1 and 6.2.2.

6.2.1 Hyperelastic Material Models

We recall from the discussion in Section 3.4 and Eqn. (3.4.27), that the thermodynamic restriction helped us to establish a relationship between stress (first Piola-Kirchhoff stress) and strain measure (Right stretch tensor), which was of the form

$$\mathbf{S} = 2\rho_0 \mathbf{F} \frac{\partial \hat{\psi}(\mathbf{C})}{\partial \mathbf{C}}. \qquad (6.2.3)$$

In rewriting Eqn. (3.4.27) in the form of Eqn. (6.2.3), we have omitted the variable θ, to focus on isothermal deformation processes.

Equation (6.2.3) indicates that it is possible to extract the stresses for an elastic material if we know an expression of an energy measure of the material. As indicated in Chapter 3, the energy measure for elastic material is defined as the strain energy density function, W (Eqn. (3.4.28)). Therefore, Eqn. (6.2.3) can be written as

$$\mathbf{S} = 2\mathbf{F} \frac{\partial \hat{W}(\mathbf{C})}{\partial \mathbf{C}}. \qquad (6.2.4)$$

All elastic materials, whose constitutive relations can be extracted from scalar energy functions, are called hyper-elastic materials (referred to also as Green elastic materials). Equation (6.2.4) can be expressed for $\hat{W}$ to be a function of different strain measures such as **U, V, B, E** and **e**. For example, we can write

$$\mathbf{S} = 2\mathbf{F} \frac{\partial \hat{W}}{\partial \mathbf{E}} \frac{\partial \mathbf{E}}{\partial \mathbf{C}} = \mathbf{F} \frac{\partial \hat{W}(\mathbf{E})}{\partial \mathbf{E}} \qquad (6.2.5)$$

In writing Eqn. (6.2.5), we have used the definition from Eqn. (3.2.30) that $\mathbf{C} = \mathbf{U}^2$. For infinitesimal deformations, the hyperelastic models manifest themselves in simple relations of the form (Exercise 6.1):

$$\sigma = \frac{\partial \hat{W}(\mathbf{e})}{\partial \mathbf{e}}. \qquad (6.2.6)$$

By comparing Eqns. (6.2.6), (4.2.5) and (4.2.2), we can observe that for a linear elastic material, $\hat{W}$ will be a quadratic function of **e**. By appropriately defining $\hat{W}$, it is possible to extract the non-linear features of any stress-strain relation.

The hyperelastic models are extracted after defining the scalar functions in terms of the scalar invariants associated with the tensor variables (Eqns. (2.3.11) and (2.3.12)). Expressing $\hat{W}$ as a function of the scalar invariants of $\mathbf{C}(I_C, II_C, III_C)$ and using chain rule, from Eqn. (6.2.4), we get

$$\hat{\mathbf{S}} = 2\mathbf{F}\frac{\partial \hat{W}(\mathbf{C})}{\partial \mathbf{C}} = 2\mathbf{F}\left\{\frac{\partial \hat{W}}{\partial I_C}\frac{\partial I_C}{\partial \mathbf{C}} + \frac{\partial \hat{W}}{\partial II_C}\frac{\partial II_C}{\partial \mathbf{C}} + \frac{\partial \hat{W}}{\partial III_C}\frac{\partial III_C}{\partial \mathbf{C}}\right\} \qquad (6.2.7)$$

Equation (6.2.7) states the general form of the constitutive relations that could be generated for a hyperelastic model. This model can be applied to any material and the associated parameters can be evaluated. This will be demonstrated for finite and small deformations in Sections 6.2.1.1 and 6.2.1.2, respectively.

6.2.2 Non-linear Hyperelastic Models for Finite Deformation

One of the most popular uses of non-linear hyperelastic models, especially for finite (large) deformations, is in rubber elasticity. Rubbers — both natural as well as artificial (normally termed as elastomers) – are characterized by being elastic and undergoing very large strains (of the order of 700%-800%). Rubberlike materials consist of extremely long molecules, which are connected to each other at joints called cross-links. The cross-link density in a rubber (or the average length of molecules between two cross-link points) is the most important factor affecting mechanical response of these materials. Fig. 6.2 illustrates the cross-linked network of rubber in undeformed and deformed conditions. When a force is applied to a rubberlike material, the polymer chains between the cross-links get stretched. Since the polymer chains between cross-links can be very long, large extensions are possible in such materials. The application of forces leads to stretching of polymer chains and reduces the disorder (possibilities of different conformations for a polymer chain) or entropy (Section 3.3.6.2). Therefore, the polymer chains have a tendency to return back to their original state, when the forces are removed (to maximize entropy). Hence, the rubber elasticity can be termed as behaviour of *entropic springs*.

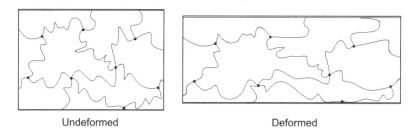

Undeformed   Deformed

Fig. 6.2 Schematic representation of crosslinked rubber in (a) undeformed state (b) deformed state

It is customary to express the strain energy density function of rubbers, in terms of the principal stretches (defined in Eqns. (3.2.32) and (3.2.33)) λ_1, λ_2

and λ_3. Since, these principal stretches are the eigenvalues of $\mathbf{U}$, we have

$$
\begin{rcases}
I_C &= \lambda_1^2 + \lambda_2^2 + \lambda_3^2 \\[4pt]
II_C &= \lambda_1^2 \lambda_2^2 + \lambda_2^2 \lambda_3^2 + \lambda_1^2 \lambda_3^2 \text{ and} \\[4pt]
III_C &= \lambda_1^2 \lambda_2^2 \lambda_3^2 .
\end{rcases}
\tag{6.2.8}
$$

Most of the rubbers are further assumed to be incompressible, so that we can use the incompressibility condition

$$
\lambda_1 \lambda_2 \lambda_3 = 1. \tag{6.2.9}
$$

Using Eqn. (6.2.9), we can modify the relations expressed in Eqn. (6.2.8) to obtain

$$
I_C = \lambda_1^2 + \lambda_2^2 + \lambda_3^2 \text{ and}
$$

$$
II_C = \frac{1}{\lambda_1^2} + \frac{1}{\lambda_2^2} + \frac{1}{\lambda_3^2}. \tag{6.2.10}
$$

The scalar invariants defined in Eqn. (6.2.10) can be used to define a general form of the strain energy density function $\hat{W}$, which is given by

$$
\hat{W}\left(I_C, II_C\right) = \sum_{i=0, j=0}^{\infty} C_{ij}\left(I_C - 3\right)^i \left(II_C - 3\right)^j. \tag{6.2.11}
$$

It should be noted that Eqn. (6.2.11) is defined for finite deformation. We note that $I_C = II_C = 3$ ($\lambda_1 = \lambda_2 = \lambda_3 = 1$) when there is no deformation. Therefore, strain energy associated with no deformation is 0. Different expressions for $\hat{W}$ can be used by choosing different number of terms from the series written in Eqn. (6.2.11). The specific forms of the strain energy function that will yield various popular models in finite deformation non-linear elasticity will be stated below.

6.2.2.1 Network Models of Rubber Elasticity

The choice of terms with $i = 1$ and $j = 0$, in Eqn. (6.2.11), will give the following expression for the strain energy density function $\hat{W}$, which is of the form

$$
W = C_{10}(I_C - 3) = C_{10}(\lambda_1^2 + \lambda_2^2 + \lambda_3^2 - 3). \tag{6.2.12}
$$

Stress-strain relations for rubbers that are derived from Eqn. (6.2.12), are known as the *network models for rubber elasticity* or *neo Hookean models*. These models were also derived by a statistical treatment of the extension of the polymer chains that constitute rubber. The model has a single material parameter (C_{10}) and this material parameter can be shown from the network theory to be directly proportional to the shear modulus of rubber as well as the absolute temperature at which the stretching is taking place. It was shown in literature that the network models are able to capture the stress-extension curves of natural rubbers very well for smaller extension ratios, but fail to capture the observed experimental behaviour near failure. Hence, the incorporation of additional terms of the series in Eqn. (6.2.11) became necessary to capture the stress-strain behaviour and this is described below.

6.2.2.2 Mooney-Rivlin Model for Rubber Elasticity

The inclusion of the term with $i = 0$ and $j = 1$, in Eqn. (6.2.11), along with the terms of Eqn. (6.2.12), will yield a strain energy density function of the following form

$$\hat{W}(I_C, II_C) = C_{10}(I_C - 3) + C_{01}(II_C - 3). \tag{6.2.13}$$

Based on the definition of invariants given in Eqn. (6.2.10), we can write

$$W = \hat{W}(\lambda_1, \lambda_2, \lambda_3) = C_{10}(\lambda_1^2 + \lambda_2^2 + \lambda_3^2 - 3) + C_{01}(\frac{1}{\lambda_1^2} + \frac{1}{\lambda_2^2} + \frac{1}{\lambda_3^2} - 3). \tag{6.2.14}$$

Stress-strain relations that are derived from Eqn. (6.2.13) are known as *Mooney models or Mooney-Rivlin models for rubber elasticity* (named after the authors who independently proposed their forms originally). This model has two independent parameters. A suitable adjustment of these two parameters was found to be sufficient to capture the stress-strain curve of natural rubber almost till its peak load. Mooney-Rivlin models also were found to perform reasonably well with most elastomers (artificial rubbers). The parameters defined in Eqn. (6.2.13) are estimated from experimental data and would have no relation with the mechanics of extension of molecular chains in rubbers. Comparing Eqn. (6.2.14) with Eqn. (6.2.12), we observe that when $C_{01} = 0$, the Mooney-Rivlin model reduces to the neo Hookean model.

Very often, the two parameters of the Mooney-Rivlin model are evaluated by fitting the parameters to experimental data from one dimensional extensional test. Assuming that the loading is in 1 direction, the symmetry of deformation implies that

$$\lambda_2 = \lambda_3. \tag{6.2.15}$$

Using Eqn. (6.2.14) and the assumption of incompressibility (Eqn. (6.2.9)), we can simplify Eqn. (6.2.13) as follows:

$$\hat{w} = C_1(\lambda_1^2 + \frac{2}{\lambda_1} - 3) + C_2(\frac{1}{\lambda_1^2} + 2\lambda_1 - 3). \tag{6.2.16}$$

When compared to Eqn. (6.2.13), we have denoted $C_{10} = C_1$ and $C_{01} = C_2$ in Eqn. (6.2.16). Using Eqn. (6.2.4), we obtain

$$S_{11} = 2F_{11}\frac{\partial \hat{W}}{\partial C_{11}} = \frac{d\hat{W}}{d\lambda_1} = 2(\lambda_1 - \frac{1}{\lambda_1^2})(C_1 + \frac{C_2}{\lambda_1}). \tag{6.2.17}$$

In Eqn. (6.2.17), we recognize that $F_{11} = \lambda_1$ and $C_{11} = \lambda_1^2$. It can be seen that stress extension ratio relationship for a neo Hookean material can be obtained by setting $C_2 = 0$ in Eqn. (6.2.17), yielding

$$S_{11} = 2C_1\left(\lambda_1 - \frac{1}{\lambda_1^2}\right). \tag{6.2.18}$$

Equation (6.2.17) can be rearranged in the following form:

$$\frac{S_{11}}{2\left(\lambda_1 - \dfrac{1}{\lambda_1^2}\right)} = C_1 + \frac{C_2}{\lambda_1}. \tag{6.2.19}$$

From Eqn. (6.2.19), we find that a plot of the left hand side with $1/\lambda_1$ would yield a straight line with a slope of C_2 and an intercept of C_1. Such a plot would be one of the easiest ways of evaluating the Mooney parameters for any rubber. However, the parameters that are evaluated in this manner will have to be verified by using the same parameters to predict the simple shear response and comparing with results from a shear test.

6.2.2.3 Ogden's Model for Rubber Elasticity

We have observed in Sections 6.2.2.1 and 6.2.2.2 that the stress-strain relations of rubbers could be captured by a single parameter as described in the network model, or by two parameters as illustrated in the Mooney-Rivlin model. A comparison of the experimental curves with the predictions of either of the models may indicate that the form of stress-strain relations as predicted by Eqns. (6.2.17) and (6.2.18), may not be adequate to capture the form of the stress-extension curves, especially at high strains.

Recognizing the limitations of the form of the equations predicted by the previous models, Ogden proposed a strain energy density function in a series form with extension ratios as

$$\breve{W} = \sum_{n=1}^{\infty} \frac{\mu_n}{\alpha_n}\left(\lambda_1^{\alpha_n} + \lambda_2^{\alpha_n} + \lambda_3^{\alpha_n} - 3\right). \tag{6.2.20}$$

Ogden proposed that the powers α_n, may take any values, positive or negative, and are not necessarily integers. The coefficients μ_n are constant coefficients that are associated with the model. It may be noted that the network model (Eqn. (6.2.12)) is a special case of the model proposed in Eqn. (6.2.20), where $\alpha_1 = 2$. Further, it may be noted that the Mooney-Rivlin model corresponds to the choice of two terms in the above model, with $\alpha_1 = 2$ and $\alpha_2 = -2$. We find that Ogden model is very versatile and can be adapted to generate stress-extension relations of a wide variety of forms. It must be noted that while it may be possible to capture all the intricacies of stress-extension relations by choosing more parameters in the Ogden model, evaluation of the associated coefficients will not be a trivial task and the modeler will have to exercise his/her own intuition in truncating the expansions suggested in Eqn. (6.2.20).

6.2.2.4 Non-linear Hyperelastic Models in Infinitesimal Deformation

We have seen in Eqn. (6.2.6) that hyperelastic models can also be used to describe non-linear elastic behaviour involving infinitesimal deformation. In component form, Eqn. (6.2.6) can be written as

$$\sigma_{ij} = \frac{\partial \hat{W}(\mathbf{e})}{\partial e_{ij}}. \tag{6.2.21}$$

The strain energy density function $\hat{W}$ defined in Eqn. (6.2.21) can be written in terms of the strain invariants I_e, II_e and III_e as

$$\hat{W} = \hat{W}(I_e, II_e, III_e). \tag{6.2.22}$$

Based on Eqns. (6.2.21) and (6.2.22), we get the following explicit stress-strain relations for a hyper-elastic model undergoing infinitesimal elastic deformation (similar to Eqn. (6.2.7)):

$$\sigma_{ij} = \frac{\partial \hat{W}}{\partial I_e}\delta_{ij} + \frac{\partial \hat{W}}{\partial II_e}e_{ij} + \frac{\partial \hat{W}}{\partial III_e}e_{im}e_{mj}, \tag{6.2.23}$$

where we note that

$$\frac{\partial I_e}{\partial e_{ij}} = \delta_{ij}; \frac{\partial II_e}{\partial e_{ij}} = e_{ij}; \frac{\partial III_e}{\partial e_{ij}} = e_{im}e_{mj}. \tag{6.2.24}$$

We can observe that the derivatives of strain energy density are related to each other in the following manner:

$$\frac{\partial}{\partial I_e}\left(\frac{\partial \hat{W}}{\partial II_e}\right) = \frac{\partial}{\partial II_e}\left(\frac{\partial \hat{W}}{\partial I_e}\right)$$

$$\frac{\partial}{\partial III_e}\left(\frac{\partial \hat{W}}{\partial I_e}\right) = \frac{\partial}{\partial I_e}\left(\frac{\partial \hat{W}}{\partial III_e}\right) \text{ and }$$

$$\frac{\partial}{\partial II_e}\left(\frac{\partial \hat{W}}{\partial III_e}\right) = \frac{\partial}{\partial III_e}\left(\frac{\partial \hat{W}}{\partial II_e}\right). \tag{6.2.25}$$

The applications of Eqn. (6.2.23) are discussed in more detail in Section (6.2.3). Another approach of arriving at non-linear elastic models of the type given in Eqn. (6.2.23) without invoking strain energy density function, is described in Section 6.2.2.

6.2.3 Cauchy Elastic Models

It is possible to expand Eqn. (6.2.1) as a series in variable $\mathbf{C}$ (similar to series in Eqns. t (4.2.1) and (4.3.1)). For many solids, the deformations are infinitesimal, and hence it is sufficient to use the infinitesimal strain measure $\mathbf{e}$ to monitor the deformation and there is no need to distinguish between the Cauchy stress σ and the Piola-Kirchhoff stress $\mathbf{S}$. In such materials, Eqn. (6.2.1) reduces to the following form:

$$\sigma = \hat{f}(\mathbf{e}) \qquad (6.2.26)$$

Consistent with the popular use of the models, we will present the Cauchy elastic models as non-linear relationships between σ and $\mathbf{e}$. Hence, the general non-linear stress strain relationship could be expressed as

$$\sigma = \beta_o \mathbf{I} + \beta_1 \mathbf{e} + \beta_2 \mathbf{e}^2 + \beta_3 \mathbf{e}^3 + ..., \qquad (6.2.27)$$

where $\beta_o, \beta_1, \beta_2,..., \beta_n$ represent response functions or coefficients. Using Cayley-Hamilton theorem (Eqn. (2.3.14)), Eqn. (6.2.27) reduces to

$$\sigma = \phi_o \mathbf{I} + \phi_1 \mathbf{e} + \phi_2 \mathbf{e}^2, \qquad (6.2.28)$$

where ϕ_o, ϕ_1 and ϕ_2 are polynomial functions of the invariants I_e, II_e and III_e. Using index notation, it is possible to rewrite Eqn. (6.2.28) as

$$\sigma_{ij} = \phi_o \delta_{ij} + \phi_1 e_{ij} + \phi_2 e_{im} e_{mj}. \qquad (6.2.29)$$

We find that Eqn. (6.2.29) has a structure that is similar to the hyper-elastic model described in Eqn. (6.2.23). In Eqn. (6.2.29), the coefficient functions ϕ_o, ϕ_1 and ϕ_2 are assumed to be independent functions. However, in Eqn. (6.2.23), the coefficients of strains are related to each other through Eqn (6.2.25). The interdependence of coefficients described in Eqn. (6.2.25), suggests that the hyper-elastic models for small deformation described in Eqn. (6.2.23) can be considered as a sub-class of the Cauchy models described in Eqn. (6.2.29). However, we observe that origins of the hyper-elastic models are clearly different from the Cauchy models. The hyper-elastic models are derived based on the assumption of the existence of a strain energy density function $\hat{w}$ associated with a material, while no such assumption is necessary for the formulation of the Cauchy models. We will now examine a few possible expansions for the polynomial functions ϕ_o, ϕ_1 and ϕ_2, which would generate the various orders of Cauchy elastic material models, that are used in engineering practice.

6.2.3.1 First Order Cauchy Elastic Models

Let us consider the following combination of polynomial functions that are to be used in Eqn. (6.2.29):

$$\phi_o = \alpha_o + \alpha_1 I_e \; ; \; \phi_1 = \alpha_2; \; \phi_2 = 0, \qquad (6.2.30)$$

where α_o, α_1 and α_2 are constant coefficients. Substituting Eqn. (6.2.30) in Eqn. (6.2.29), we obtain the expression for first order Cauchy elastic model as

$$\sigma_{ij} = (\alpha_o + \alpha_1 I_e) \delta_{ij} + \alpha_2 e_{ij}. \qquad (6.2.31)$$

Equation (6.2.31) shows a linear relationship between stresses and strains, and hence is equivalent to the isotropic linear elastic relations that are discussed in Chapter 4 (Eqn. (4.2.1)). We also observe that the first coefficient α_0 is not multiplied with any strain and hence could be treated as a coefficient that will represent the initial or reference stress in the material. It can be proven through simple calculations (see Exercise 6.3) that the coefficients α_1 and α_2 can be related to the Lame's coefficients λ and μ that have been defined for isotropic elastic models (Eqn. (3.2.15)).

6.2.3.2 Second Order Cauchy Elastic Models

In order to generate a quadratic expansion for the stresses in terms of the strains, we would have to judiciously choose the polynomial expansions for the functions ϕ_0, ϕ_1 and ϕ_2 in such a way that coefficients up to the quadratic expansion for the stresses are included in the specification of these polynomials. Hence, it would be necessary that the polynomials are of the following form:

$$\left. \begin{array}{l} \phi_0 = \alpha_1 I_e + \alpha_2 I_e^{\,2} + \alpha_3 II_e \\ \phi_1 = \alpha_4 + \alpha_5 I_e \text{ and} \\ \phi_2 = \alpha_6 \end{array} \right\} \qquad (6.2.32)$$

Substituting the polynomial functions from Eqn. (6.2.32) into Eqn. (6.2.29), we obtain the following form for the second order Cauchy elastic model:

$$\sigma_{ij} \cdot = \left(\alpha_1 I_e + \alpha_2 I_e^{\,2} + \alpha_3 II_e \right) \delta_{ij} + \left(\alpha_4 + \alpha_5 I_e \right) e_{ij} + \alpha_6 e_{im} e_{mj}. \qquad (6.2.33)$$

We note that a general expansion for the second order Cauchy model as given in Eqn. (6.2.33) could have at the most six independent coefficients (ignoring the coefficient associated with the initial stress, that was considered in the first order Cauchy models). The six coefficients in Eqn. (6.2.33) will have to be evaluated by observing the material response in independent mechanical tests such as simple tension and simple shear tests.

6.2.4 Use of Non-linear Elastic Models

Since Eqn. (6.2.1) has $\mathbf{C}$ as the independent variable, it can be visualized as an expression describing displacement controlled deformation of material. Similarly, Eqn. (6.2.2) is an expression where $\mathbf{S}$ is an independent variable and hence it can be visualized as a load control deformation in material. In literature, the term *non-linear elastic material* is also used by allowing the loading to take place through only a deformation control (constitutive relations of the form given in Eqn. (6.2.1)) and ignoring unloading. While we recognize that such material behaviour and its modeling are not *elastic*. However, to describe this behaviour, non-linear elastic models described in Sections 6.2.1 and 6.2.2 are used widely in engineering practice. Soils, concrete and biological tissues are some examples of such materials.

6.3 NON-LINEAR INELASTIC MODELS

In Section 6.2 above, we have discussed the *non-linear elastic* models. The elasticity of the models are primarily characterized by the following properties:

(*a*) The materials retrace their loading path and return to their original state of deformation, after the load causing deformation is removed.

(*b*) The state of stress in the material, is purely a function of the state of strain in the material and not on the path followed by the material to reach a given state.

The assumptions stated above, normally also ensure that there is a unique inverse relationship between the strain and the stress in a material. The models described in Section 6.2 are ideally suited for materials where there is no internal dissipation (Section 3.3.6.2) in a material. Some materials like rubbers and elastomers, would exhibit this reversible stress-strain response right upto their failure loads. Many other materials like concrete, soils, metals etc. would show a non-linear reversible response upto a critical load (normally called as the yield load), beyond which dissipative processes start influencing the material behaviour. These materials show non-linearity in their stress-strain response, but this response is not totally reversible. This would mean that the material will not return to its original state of stress, when the load causing the deformation is removed. Further, the equilibrium state (indicating a correlation between stress and strain) is a function of the loading history or the path followed by the material to reach a particular state. Such materials can be classified as *non-linear inelastic* materials.

There are several ways of modelling the non-linear inelastic response of materials. One of the classical ways of modelling inelastic phenomenon is to identify the departure of the material from elastic behavior in its stress-strain curve. This point is termed as *yielding* of the material. The mechanical response of the material beyond the yield point is identified as a *hardening* or a *softening* phenomenon and some postulates are made to describe this phenomenon. A treatment of inelastic behaviour in this manner forms a part of *plasticity theory* and will be described in more detail in Section 6.4 later in this chapter.

In many materials like soils, it may be difficult to identify a distinct point like a yield point, at which departure from elastic behavior takes place. In such materials, the inelastic behaviour is assumed to manifest right from the origin of the stress-strain curve. The use of plasticity theory is found to be very difficult to model the mechanical response of such materials. It is found to be convenient to monitor the evolution of inelastic behavior in such materials through an internal state variable called as a *damage variable*. The use of damage variable (also known as *degradation parameter*) and its evolution with loading, in the modelling of stress-strain response of materials will be illustrated briefly through a case study in Section 6.5 later in this chapter. A path dependent inelastic behaviour may also be modelled as an agglomeration of *incremental linear behaviour*, where the linear behaviour is assumed to manifest within each increment of loading. An inelastic behaviour of this type is termed as a *hypoelastic* model and will be described briefly in the Section 6.3.1.

6.3.1 Hypoelastic Material Models

In a hypoelastic material, the stress rate depends on current level of stress and a measure of rate of deformation. Therefore, a general relationship for a constitutive model of a hypoelastic material will be of the form:

$$\dot{\sigma}_{ij} = \hat{f}\left(\sigma_{kl}, D_{kl}\right).$$
(6.3.1)

For small strains, Eqn. (6.3.1) reduces to the following form:

$$\dot{\sigma}_{ij} = \hat{f}\left(\sigma_{kl}, \dot{e}_{kl}\right).$$
(6.3.2)

It can be shown that all isotropic functions that are defined to be of the form Eqn. (6.3.2), can be expanded to have a general form which is given below:

$$\left. \begin{aligned}
\dot{\sigma}_{ij} &= \left(\alpha_o \dot{e}_{kk} + \alpha_1 \sigma_{mn} \dot{e}_{nm} + \alpha_2 \sigma_{mn} \sigma_{nk} \dot{e}_{km}\right)\delta_{ij} \\
&+ \left(\alpha_3 \dot{e}_{kk} + \alpha_4 \sigma_{mn} \dot{e}_{nm} + \alpha_5 \sigma_{mn} \sigma_{nk} \dot{e}_{km}\right)\sigma_{ij} \\
&+ \left(\alpha_6 \dot{e}_{kk} + \alpha_7 \sigma_{mn} \dot{e}_{nm} + \alpha_8 \sigma_{mn} \sigma_{nk} \dot{e}_{km}\right)\sigma_{im}\sigma_{mj} \\
&+ \alpha_9 \dot{e}_{ij} + \alpha_{10}\left(\sigma_{im}\dot{e}_{mj} + \dot{e}_{im}\sigma_{mj}\right) + \alpha_{11}\left(\sigma_{im}\sigma_{mk}\dot{e}_{kj} + \dot{e}_{im}\sigma_{mk}\sigma_{kj}\right).
\end{aligned} \right\}$$
(6.3.3)

In Eqn. (6.3.3), the coefficients $\alpha_0, \dots, \alpha_{11}$ are functions of invariants of stress.

As described earlier, a hypoelastic model is visualized as *an elastic model for small changes in strains and stresses*. Therefore, an elastic behaviour is assumed to exist only within a small incremental region of the total mechanical response. The mechanical response of such a material within these incremental loading, is stress dependent and will depend upon the state of stress existing in the body at the stage of load increment. Since the increments in stress and strain in such incremental models are homogeneous with time, it is conventional to consider the incremental stresses $d\sigma$ as equivalent to stress rates $\dot{\sigma}$. Similarly, the incremental strains are considered equivalent to strain rates $\dot{\varepsilon}$. With these modifications, the constitutive relations for hypoelastic models will look very similar to the relations developed for rate dependent viscoelastic models. However, it must be remembered that the rates implied here do not have any inertia terms associated with them.

Such incremental formulation of hypoelastic models is useful for engineering materials such as soils, other granular materials and geomaterials. Since these models describe the material response in small increments, they are defined in an infinitesimal sense using Eqn. (6.3.2). Equation (6.3.3) can also be written as a constitutive relation of the form:

$$\dot{\sigma}_{ij} = C_{ijkl}\left(\sigma_{mn}\right)\dot{e}_{kl}.$$
(6.3.4)

Equation (6.3.4) states a linear relationship between stress rates and strain rates. The associated incremental equations can be written as

$$d\sigma_{ij} = C_{ijkl}\left(\sigma_{mn}\right)de_{kl}. \tag{6.3.5}$$

It is possible for us to construct hypoelastic constitutive relations in accordance to Eqn. (6.3.5), for specific loadings. If necessary, the incremental relations could be integrated to obtain an explicit relation between stress and strain. The number of parameters that need to be evaluated in such a relation will depend upon the loading conditions. This is illustrated in Example 6.3.1.

Example 6.3.1: A grade one hypoelastic model is defined in the following way:

$$\alpha_0 = a_0 + a_2\sigma_{nn}; \ \alpha_9 = a_1 + a_3\sigma_{nn};$$

$$\alpha_3 = a_4; \ \alpha_2 = a_6; \ \alpha_{10} = a_5; \text{ and all other } \alpha_i\text{s are zeros.}$$

Derive the grade one hypoelastic model for hydrostatic loading of this material.

Solution:

A hydrostatic loading is characterized by the following states of stress:

$$\sigma_{11} = \sigma_{22} = \sigma_{33} = \sigma; \ \sigma_{12} = \sigma_{23} = \sigma_{13} = 0;$$

$$e_{11} = e_{22} = e_{33} = e; \ e_{12} = e_{23} = e_{13} = 0. \tag{E6.3.1}$$

Substituting the grade one model and the stress/strain state of hydrostatic loading in Eqn. (6.3.5, we get the following relationship between the incremental stresses and strains, which is of the form:

$$\frac{d\sigma}{c_1 + c_2\sigma} = de, \tag{E6.3.2}$$

where c_1 and c_2 are arbitrary constants. Assuming that the initial strain is zero and that the initial stress is σ_0, it is possible to integrate the above equation to obtain the stress strain relation for this hypo-elastic model to be of the form:

$$\sigma = \left(\frac{c_1}{c_2} + \sigma_0\right)\exp(c_2 e) - \frac{c_1}{c_2}. \tag{E6.3.3}$$

6.4 PLASTICITY MODELS

It was stated in Section 3.2 that there are *inelastic* solid like materials (exemplified by curve (c) in Fig, 3.1), which do not retrace their stress-strain trajectory, when they are unloaded at certain stages of their deformation history. The inability of the material to retrace its loading path was attributed to be due to dissipative mechanisms in the material. These dissipative mechanisms result in *plastic deformation*, which are observed in metals (predominantly at high temperatures), soils and granular materials, and brittle materials. The behaviour and modelling of such materials is discussed briefly in the current section.

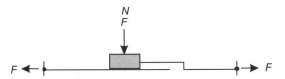

Fig. 6.3 Friction analogy for the onset of plastic deformation

Plastic deformations in a material can be visualized to be triggered when the state of stress in the material reaches a *critical condition*. This sudden onset of plastic deformation, is similar to the sliding of a block on a rough surface due to the application of a pulling or a pushing force F. This concept is illustrated schematically in Fig. 6.3. We know from Coulomb's experiments that motion takes place in this block only when the applied force F reaches a critical value F_s given by

$$F_s = \mu_s N, \tag{6.4.1}$$

where μ_s is the coefficient of static friction between the sliding surfaces and N is the normal force acting on the sliding block. On similar lines, plastic deformation takes place in a material when some critical conditions are experienced in the material, after an initial elastic deformation. Hence, the onset of plastic deformation is often treated as a sudden (or a singular event) that is different from other types of deformations that normally occur in the material. It may be noted that plastic deformations plastically deforming materials, are special type of deformations, that are different from elastic deformations or viscous deformations. Unlike elastic deformations, there is a flow involved in plastic deformation and unlike fluids, this flow is not explicitly dependent on the rate of strain.

Plastic deformations have conventionally been attributed to a process of slip in metallic crystals. In such deformations, there is no significant separation of the material taking place though the existing surfaces may extend or contract in a continuous sense. Hence, plastic deformation is conventionally classified as a continuum deformation. There is a qualitative match in the deformation behaviour at a macroscopic scale between the metallic materials and some of the other common engineering materials such as concrete, clay, rock, etc. In these materials, inelastic deformations are associated with void growth, formation of microcracks, granular rearrangement etc. The macroscopic effects of these processes are similar to the macroscopic effects seen in the deformation of metals. Hence, these materials are also modelled along the lines of metal plasticity theory, by including certain additional conditions such as pressure dependent yielding etc. In many applications, the material is only subjected to a monotonically increasing load during its life time. Plastic deformations in such applications are often captured through *monotonic plastic deformation* models. In applications where the material is subjected to unloading and reverse loading, the plastic deformations are captured incrementally through *incremental plastic deformation* models. Both these behaviours will be described in detail in this section.

In Section 6.4.1, we will examine a typical macroscopic uniaxial response of a material undergoing plastic deformation. In Section 6.4.2, we will examine the general modelling assumptions and the popular *monotonic plastic deformation* models of plasticity. In Section 6.4.3, we will examine the use of *incremental plastic deformation* models in capturing the same phenomenon that was described in Section 6.4. Finally, in Section 6.4.4, we will examine note some special features that will manifest themselves when the material is subjected to cyclic loading.

6.4.1 Typical Response of a Plastically Deforming Material

It would be useful to examine a typical response of a plastically deforming material, when it is subjected to monotonically increasing loads as well as to cyclic loads. Fig. 6.4 shows typical response of an engineering material when it is subjected to a uniaxial tension/compression test. The following observations may be made with regard to the different regions in the figure.

1. *Elastic Response*: Up to a certain critical stress (which is normally the initial yield stress $\sigma_y^{\,i}$), the stress is linearly proportional to strain. If the material is unloaded before it reaches $\sigma_y^{\,i}$, the strain is fully recovered and the plot shows a complete trace back to the origin. It may be noted that this initial elastic response may even be insignificant in many materials (such as a soils), so that plastic response (to be described in stages 2–5 below), may be postulated to be occur even at very low loads in such materials.

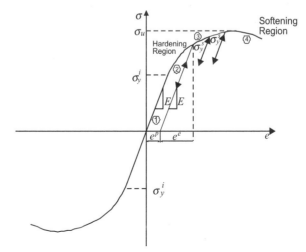

Fig. 6.4 Typical stress-strain plot of a plastically deforming material

2. *Yield Point:* There is a point (σ_y^i) on the stress strain plot (though a little difficult to locate experimentally), at which the linear reversible behaviour described in stage 1 ends and there is an excessive build up of strains without a proportional increase in stresses in the material. The friction block analogy, signifying the onset of plastic deformation, as described in Fig. 6.3, can be postulated to be activated at this stage. It may be noted that there is usually a degradation in modulus that is seen in the material, when it is loaded beyond σ_y^i. Further, as can be seen in Fig. 6.4, the material does not retrace the loading path when unloaded at stress that are greater than σ_y^i. The transition of any material from stage 1 to stage 2 is often so smooth that it is hard to detect a single point of transition in the slope (modulus) of the stress-strain curve.

3. *Residual Strain* : The next stage is a stage of unloading when the stresses are larger than σ_y^i. We notice that the unloading curve does not retrace the loading path. Hence, there is a *residual strain* (or permanent strain) e^p at zero load. This residual strain is also sometimes called as *plastic strain* associated with the loading cycle. As can be observed in the figure, the unloading path is invariably *linear* and is referred to as *elastic unloading.*

4. *Plastic Flow*: An accumulation of residual strains in a material is called as *plastic flow* in the material and the yield point is normally considered as the stress state that marks the *onset of plastic flow* in the material. We note here that, in general, the unloading curve has a slope equal to that of the slope of the initial proportional loading (stage 1).

5. *Hardening*: When the unloaded material (with a residual strain of e^p) is reloaded, the material typically shows a linear elastic response until the stresses reach another yield point, which we call as *subsequent yield point, σ_y^s.* When loaded beyond σ_y^s, plastic flow starts again in the material. For the material to experience a stress state of σ_y^s (which is greater than σ_y^i), plastic flow must have occurred in the material. This process of realizing higher yield points in the material due to plastic flow, is called as *hardening*.

6. *Reverse Loading*: If loading is applied in the negative direction (compression), such loading is called as *reverse loading*. It may be noted that in many materials (metals), the compressive yield point is the same as the tensile yield point, while in quite a few materials (soils and concrete), the compressive yield point may be different from the tensile yield point. This phenomenon of lowering of yield stress upon reverse loading is called the *Bauchinger effect*. These differences in the yield points in compression and tension form the basis for different hardening models as will be explained in Section 6.4.4 below.

7. *Softening*: Plastic deformation persists in the deformational response of the material till the resistance to loading reduces at σ_u. This maximum stress carried by the material σ_u is called as *ultimate strength* of the material. When loading is continued beyond σ_u, one finds a neck formation at an arbitrary location within the gage length of a tensile test specimen in metals. The size of the neck continues to decrease and the load carried by the specimen reduces, till complete separation occurs in the specimen. The response of the material beyond σ_u is called as *softening*.

Keeping in view the basic features of plastic deformation described in the steps 1–7 above, we will now discuss the mathematical modelling of each of these stages briefly in the next section.

6.4.2 Models for Monotonic Plastic Deformation

As mentioned in the introduction earlier, a plastically deforming material can be loaded *monotonically* up to ultimate load. Monotonic loading would imply a continuous increase of load/strain in the material until failure. This would imply that the material will experience the states of initial yield σ_y^i and all subsequent yield points σ_y^s even without being subjected to a load reversal, until the material reaches the ultimate load σ_u. Hence, modelling for monotonic plastic deformation would involve the development of an appropriate expression that would capture the non-linear response of the material beyond the initial yield σ_y^i. While developing such models, the total stresses and strains will be considered (and not incremental stresses/strains as will be dealt later with in incremental models in Section 4.5.3). The following assumptions are inherently made while developing expressions in these models

(*a*) The material is linear elastic until it reaches the initial yield σ_y^i. Hence, stresses and strains are related in this region through the expression

$$\sigma = Ee. \tag{6.4.2}$$

where E is the tensile modulus in the material is the strain (within the elastic region).

(*b*) The total plastic strain e in the material beyond the initial yield point is visualized to be additive combination of an elastic strain e^e and a plastic strain e^p so that it is possible for us to write

$$e = e^e + e^p. \tag{6.4.3}$$

It may be noted that a more generalized form of the additive decomposition shown above, is multiplicative decomposition of deformation gradient $\mathbf{F}$ in terms of the elastic deformation gradient $\mathbf{F}^e$ and a plastic deformation gradient $\mathbf{F}^p$ of the type

$$\mathbf{F} = \mathbf{F}^e\mathbf{F}^p. \tag{6.4.4}$$

An equivalence between Eqns. (6.4.3) and (6.4.4), for small strains is left as an exercise to the reader (see Exercise 6.6).

Tabel 6.1 List of Existing Models for Deformational Plasticity

Model	Remarks	Equations	Response Modelled
Rigid perfectly plastic	(a) Elastic strains are absent. (b) No hardening	$\sigma_y = \sigma_y^j$ $e = e^p$	
Rigid linear hardening	(a) Elastic strains are absent. (b) Linear hardening	$\sigma_y = c e^p$ $e = e^p$	
Rigid power law hardening	(a) Elastic strains are absent. (b) Power law hardening	$\sigma_y = c\left(e^p\right)^n$ $e = e^p$	

Elastic Perfectly Plastic		$\sigma = \begin{cases} Ee, e<e_y \\ \sigma_y, e \geq e_y \end{cases}$ with $e_y = \sigma_y/E$ $e = e_y + e^p$	(a) Elastic strains are present. (b) No hardening
Elastic linear hardening		$\sigma = \begin{cases} Ee, e\leq e_y \\ Ee_y + c(e-e_y), e\geq e_y \end{cases}$ and $c < E$ $e = e_y + e^p$	(a) Elastic strains are present. (b) Linear hardening
Elastic exponential Hardening		$\sigma = \begin{cases} Ee, e<e_y \\ Ee_y + k(e-e_y)^n, e\geq e_y \end{cases}$ $e = e_y + e^p$	(a) Elastic strains are present. (b) Exponential hardening (c) Hardening parameters k and n related to each other (due to continuity at e_y)

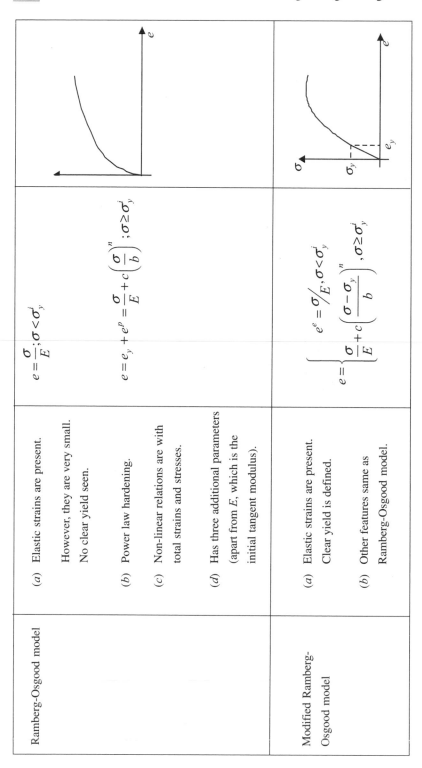

Ramberg-Osgood model	(a) Elastic strains are present. However, they are very small. No clear yield seen. (b) Power law hardening. (c) Non-linear relations are with total strains and stresses. (d) Has three additional parameters (apart from E, which is the initial tangent modulus).	$e = \dfrac{\sigma}{E}; \sigma < \sigma_y^i$ $e = e_y + e^p = \dfrac{\sigma}{E} + c\left(\dfrac{\sigma}{b}\right)^n ; \sigma \geq \sigma_y$
Modified Ramberg-Osgood model	(a) Elastic strains are present. Clear yield is defined. (b) Other features same as Ramberg-Osgood model.	$e = \begin{cases} e^e = \sigma/E, \sigma < \sigma_y^i \\[2mm] \dfrac{\sigma}{E} + c\left(\dfrac{\sigma - \sigma_y}{b}\right)^n, \sigma \geq \sigma_y \end{cases}$

Table 6.2 Choice of Appropriate Plasticity Models for Various Applications

S. No.	Application	Features	Model
1.	Metal forming applications such as metal extrusion, wire drawing, sheet drawing, punching etc.	Large plastic strains and small elastic strains	Rigid-plastic assumption to simulating the process for initial die design (simple deformational theory if single pass)
2.	Spring back upon extrusion etc.	Large plastic strains and small elastic strains	Elastic-plastic assumption with additive decomposition of elastic and plastic strains
3.	Very low cycle fatigue	Finite plastic strains and comparable elastic strains	Finite deformational plasticity — multiplicative decomposition of strains
4.	Low cycle fatigue	Small plastic strains and elastic strains	Small deformation cyclic plasticity (incremental plasticity, (Section 6.4.3 described below))
5.	Inducing residual stresses such as shot peening, etc.	Small plastic strains and elastic strains	Small deformation theory of plasticity
6.	Limit load design	With or without elastic together with perfectly plastic assumption	Small deformation theory of plasticity

Modelling of stress-strain relations involving total stresses and total strains for a plastically deforming material, is part of an approach in plasticity which is called as *deformational theory of plasticity*. Models based on this theory have been primarily developed for monotonic loading in the material. However, these models are often used to capture yield until reverse loading occurs, even in reverse/cyclic loading. In Table 6.1, we summarize some of the popular available models from the literature, along with a brief description of each of the models. It may be noted that the table consists of two broad classes of modes, *viz.*, (*i*) rigid plastic models and (*ii*) elastic-plastic models. Rigid plastic models are primarily find applications in metal forming while the elastic-plastic models find application in other structural applications. In Table 6.2, some common applications for which plasticity is to be studied, the features to be modelled and the appropriate models suitable for modelling are listed. This can be used as a general guideline for choosing an appropriate model.

6.4.3 Models for Incremental Plastic Deformation

The models described in Section 6.4.1, will be capable of capturing the total non-linear response of the material, if it is uniquely defined. If the material response has multiple non-linear paths, each of which is not clearly defined, the monotonic models described in Section 6.4.1 will be difficult to develop and use. Multiple non-linear paths do occur when the material is subjected to repeated loads and such situations do occur in many applications. To model such loading situations, there is a need to develop models that can capture and accumulate the effects of plastic deformations in an incremental fashion. The *incremental plastic deformation models* address these needs by developing relations that help to capture the incremental stress-strain relationship instead of the total stress-total strain relationship. This incremental theory can then be used to model any complicated loading history (including cyclic loadings) that the material may undergo during its loading history.

It is important to note that in an incremental formulation, we are monitoring the response of the material due to incremental loads. Thus, in a stress-strain plot, we are observing the incremental response de^j due to an incremental stress $d\sigma^j$. It is to be noted that in an incremental formulation, the state of stress/strain is known in the material upto the previous load step. In other words, the state of stress (σ^{j-1}) or strain (e^{j-1}) is known upto the step $j - 1$. We are seeking the total stress (σ^j) or strain (e^j) in the material due current increments $d\sigma^j$ and de^j through the relations

$$\sigma^j = \sigma^{j-1} + d\sigma^j$$

and $$e^j = e^{j-1} + de^j. \tag{6.4.5}$$

In incremental plasticity, the primary focus is to develop a consistent procedure that will help us to estimate the instantaneous secant modulus E_{eff}^j, so that it is possible to establish a relationship of the following form

$$d\sigma^j = E^j_{eff} de^j .$$ (6.4.6)

Various models in incremental plasticity are effective ways by which E^j_{eff} is obtained for a given material. It may be noted that E^j_{eff} will be *positive* when we are tracing the hardening portion of the curve in Fig. 6.4 and it will be *negative* when we are tracing the softening part of the response. It may also be noted that Eqn. (6.4.5) helps us to proceed from one state of equilibrium to another state of equilibrium associated with the material response. The *loading* is associated with accumulation of plastic strains and an associated increase in elastic strains. When the loads are reduced from any given yield point, the material retraces an elastic path as indicated in Fig. 6.4. Hence, any model in incremental plasticity that attempts to capture the loading path of a material, should be able to identify the following distinct phases in the material behaviour

(*a*) Instantaneous yield condition in the material.

(*b*) Criteria for loading or unloading in the material.

(*c*) Estimation of plastic strain in the material.

(*d*) Estimation of increase in total stress associated with the loading.

Each of the items mentioned above will be explained in detail in the following description:

(*a*) *Instantaneous yield condition*: The yield conditions for various models for deformational plasticity have been already listed in Table 6.1. A simple definition of yield can then be given by the expression

$$|\sigma| = \sigma^j_y, j = i, s; s = 1, 2, 3, ...,$$ (6.4.7)

where σ^j_y refers to the *initial* as well as all *subsequent yields* in the material, as indicated in Fig. 6.4. It is to be noted that the yield criteria is defined with reference to an absolute value of stress $|\sigma|$, so that this definition accounts for load reversals that are likely to take place in the material. The criteria for yield defined in Eqn. (6.4.7) can also be defined in terms of a *yield function f* defined as

$$f := |\sigma| - \sigma^j_y = 0; j = i, s; s = 1, 2, 3, ...$$ (6.4.8)

It is to be noted that *f* will take a value of *zero* once yield occurs in a material and will be negative when yielding has not yet occured. The definition of yield as observed in Eqn. (6.4.8), will continue to be valid in a multi-axial state of loading also, with a pre-defined scalar measure σ^e known as the effective stress will replace $|\sigma|$ in Eqn. (6.4.8). The effective stress σ^e is normally defined in terms of the stress invariants. An example of effective stress and its use is given in Exercise 6.7 .

(b) *Criteria for loading or unloading in the material:* Broadly, we all understand *loading* to be a situation where applied stresses increased in the material and *unloading* as a situation when the applied stresses are decreased in the material. In a stress strain curve, loading is associated with an *increase* in the stresses and their strains, and unloading is associated with a *decrease* in the same quantities. These are indicated by appropriate arrows in the representative stress-strain curves shown in Fig. 6.4. It may be noted in Fig. 6.4 that unloading is invariably *elastic* while loading is *elastic* till we reach the yield point and *plastic* thereafter. The yield function f defined in Eqn. (6.4.8) can be used to define yield, while an increment in this function df can be used to define the state of loading or unloading in the material. Hence, it is possible to write

$$\left. \begin{array}{l} f = 0, df > 0 \rightarrow \text{plastic loading,} \\ f = 0, df < 0 \rightarrow \text{elastic unloading} \qquad \text{and} \\ f = 0, df = 0 \rightarrow \text{neutral loading.} \end{array} \right\} \qquad (6.4.9)$$

Equation (6.4.9) define the *loading criteria* in a plastically deforming material. It may be noted the above equation also lists a state of neutral loading, where there is neither an accumulation of plastic strain nor a decrease in the elastic stresses/strains. The significance of neutral loading will be understood well, if we are able to describe mathematically the other state variable Q, which along with the applied stress σ indicates yield in a material. Therefore it is possible to write a simple representation of the yield function as

$$f = f(\sigma, Q). \qquad (6.4.10)$$

The nature of this state variable may differ with each material as well as the assumptions that are made in a model. For example, the state variable may be capturing the dislocation density in a metal, or air void density in a brittle material. Several state variables may be associated with the definition of a yield function f. The state variable can also be another scalar function that represents additional variables that are responsible for postulated mechanisms of plastic deformation. In cyclic plasticity (to be described later in Section 6.4.4 below), we use two scalar functions α and R each of which is further a function of the total plastic strain e^p in the material. These concepts will be explained later in Section 6.4.4. It is to be noted that the scalar functions or the scalar variables will evolve (grow) with loading and hence, an evolution equation along with an appropriate initial conditions, will become an essential features associated with the formulation of a model in incremental plasticity. The stresses could also be in the form of a scalar stress invariant representing all the three principal stresses that exist in a material. A neutral loading will signify a movement on the yield surface, where there is an internal rearrangement between the various stresses or state variables and *yet*, there is no increase in the yield strain in the material. For simplicity, we will be now developing the concept of neutral loading using a single scalar stress variable σ and a single state variable Q. Substituting the parametric description of f from Eqn. (6.4.10) into the definition of neutral loading defined in Eqn. (6.4.9), and performing the perturbation, it is possible to express the conditions for neutral loading as

$$df\left(\sigma,Q\right)=0 \Rightarrow \frac{\partial f}{\partial \sigma}d\sigma + \frac{\partial f}{\partial Q}dQ = 0. \tag{6.4.11}$$

Equation (6.4.11) is also called as the *consistency condition* and gives us a relation between the stress increments *ds* and incremental changes in the state variable *dQ*. These relations will be useful in estimating the effective modulus E_{eff}^{j} as will be illustrated in Example 6.4.1 later in this section.

(c) *Estimation of plastic strain in the material:* The estimation of incremental plastic strains de^{p} associated with every increment in loads $d\sigma$. Earlier, while dealing with total deformational models, we have decomposed the total strains into elastic strains and plastic strains as per Eqn. (6.4.3). An incremental equivalent of the same equation can be written as

$$de = de^{e} + de^{p}, \tag{6.4.12}$$

where the total increment in plastic strain (the superscript *j* introduced in Eqn. (6.4.6), which was monitoring the stage of loading is dropped for simplicity) is an elastic increment de^{e} and a plastic increment de^{p}. The elastic increment de^{e} can be related to the incremental stresses *ds* through a relation treated as an extension of Eqn. (6.4.5), so that we can write

$$d\sigma = E_{ul}\, de^{e}, \tag{6.4.13}$$

where E_{ul} is the unloading modulus of the material. While the unloading modulus E_{ul} can in general be different from the initial modulus E, in most simplistic models, they are assumed to be the same. An estimation of the plastic flow increment is addressed through a *flow rule* in plasticity, which will be explained briefly here. Accumulation of plastic strain in a material at any stage of loading follows the friction block analogy that was introduced in Eqn. (6.4.1). The sliding of a friction block, in the friction block analogy, takes place at a critical loading stage and occurs in the same direction as the applied load. In a similar fashion, plastic strain accumulation in a plastically deforming material, will take place in the direction of the applied driving force only. The application of the applied force, is often defined with reference to the yield function f defined in Eqn. (6.4.8). For a given yield point, we find that there is a one to one correspondence between the yield function f and the applied stress σ, in one dimensional loading. Hence, the direction of loading will be in the same direction as f, and will have a magnitude that is proportional to f. In a multi-axial loading situation, where f represents a yield surface, loading is normally considered to be normal to the yield surface and hence the incremental plastic strain de^p will be in the direction of the yield surface. This statement of a normality of incremental plastic strain, with reference to the yield function, is presented as a *flow rule* in the theory of plasticity. Mathematically, the flow rule is represented as

$$de^p = d\lambda \frac{\partial f}{\partial \sigma},\tag{6.4.14}$$

where $d\lambda$ is plastic multiplier constant that captures the magnitude of plastic strain increment, associated with a material, for given increment in loads. The plastic multiplier λ is related to the state variable Q defined in Eqn. (6.4.11) and these relations will be described in greater details in the following discussion.

(d) *Estimation of increase in total stress associated with the loading*: The total increment in the load ds is estimated elastic load increments de^e using the unloading modulus E_{ul}. Hence, by combining Eqns. (6.4.11), (6.4.12) and (6.4.14), it is possible to write the following expression.

$$d\sigma = E_{ul}\, de^e = E_{ul}\left(de - de^p\right) = E_{ul}\left(de - d\lambda\frac{\partial f}{\partial \sigma}\right)\tag{6.4.15}$$

Equation (6.4.15) gives us the frame for deriving the effective modulus E_{eff}^j defined in Eqn. (6.4.6), provided we know the yield functions f and the plastic multiplier $d\lambda$ in the material. Obtaining these Functions and thereby the estimation of E_{eff}^j will be demonstrated in Example 6.4.1 later in this chapter.

6.4.4 Material Response Under Cyclic Loading

The total deformational models of Section 6.4.1 and the incremental deformation models of Section 6.4.3, were mathematical tools developed to capture the overall material response described in Section 6.4.1. While describing the material response in Section 6.4.1, the focus was primarily on the material response for single cycle of loading and unloading. However, it is to be noted that materials in many applications are subjected to a cyclic state of loading and it is imperative to capture the material response, when the material is subjected to cyclic loads. It is often observed that cyclic loading will induce a phenomenon called *fatigue* in the material, which is essentially the lowering of strength of the material due to repeated loading, well below the ultimate strength of the material. It is generally postulated that the accumulation of plastic strain, plays a very important role in determining the fatigue behaviour of the material.

Even though tools developed in sections 6.4.1 and 6.4.2 can directly be translated to be used when the material is used for cyclic loads, some special response features are noticed in many materials when they are subjected to cyclic loads. These response features are:

(*a*) Cyclic hardening

(*b*) Ratcheting and shakedown.

Each of these special features will be described briefly in this section.

(*a*) *Cyclic hardening*: Cyclic hardening can broadly be defined as a hardening that is induced in the material due to cyclic loading. Hardening induced due to cyclic loading is schematically represented for uniaxial loading in Fig. 6.5.

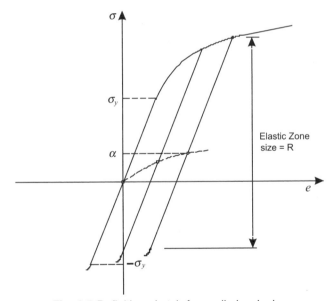

Fig. 6.5 Definition sketch for cyclic hardening

As noted in Fig. 6.5, cyclic hardening is characterized by the existence of two features. These are

(*i*) *Translation of the elastic region*: We note that there is a possibility of translation of the elastic region of stress-strain plot, without a change in the magnitude of the elastic straining in the material. This feature is called as *kinematic hardening* and is characterized by the parameter a in Fig. 6.5.

(ii) Expansion of the elastic region: We note that the total elastic region can also increase in magnitude due to cyclic loading. This type of hardening is called as *isotropic hardening* and is denoted by the parameter R in Fig. 6.5.

The translation and expansion of the yield surface indicated for uniaxial behaviour in Fig. 6.5, can even be visualized in the multi-axial stress plane as illustrated in Fig. 6.6. In this figure, the yield locus is indicated by a circle. *Isotropic hardening* is associated with an expansion of this yield surface and is indicated in Fig. 6.6(*a*). *Kinematic hardening* is associated with a translation of this yield surface and is indicated in Fig. 6.6(*b*).

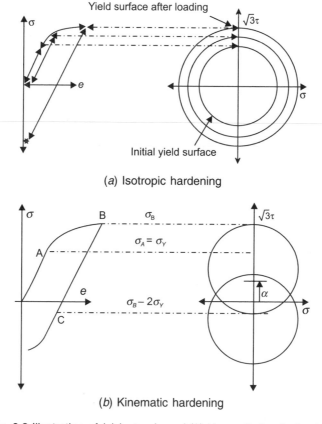

(*a*) Isotropic hardening

(*b*) Kinematic hardening

Fig. 6.6 Illustration of (*a*) isotropic and (*b*) kinematic hardening in a stress strain diagram and in a principal stress diagram

The cyclic hardening parameters a and R are normally sufficient to capture all the basic features that are associated with cyclic hardening in a material. They are normally treated as the two significant state variables (Qs of Eqn. (6.4.10), while developing incremental models in cyclic plasticity. These parameters essentially register the effect of the loading history on the material, and this loading history is likely to influence the future response of the material to cyclic loads. Hence, capturing the evolution of the hardening parameters is an important step in the modelling of material response to cyclic loading. In many models, the evolution of the hardening parameters is monitored through the evolution of *accumulated plastic strain* e_p in the material which is defined as

$$e_p = \int_{\substack{loading \\ path}} \left| de^p \right|. \qquad (6.4.16)$$

In some models, the *plastic work* w_p done by the loads defined by

$$w_p = \int_{\substack{loading \\ path}} \sigma de^p \qquad (6.4.17)$$

is used as a variable that characterizes the hardening parameters. It is to be noted that both e_p and w_p are variables that satisfy the twin properties of (*i*) growing with loading and (*ii*) positive in nature. Any other variable satisfies these twin properties could also be chosen to be the defining variable for the hardening parameters. The specific form of hardening parameters, their use in modelling of plastic deformation and their evolution, will be illustrated in Example 6.4.1.

A special case of cyclic hardening, is the material associated with constant *stress cycling*. The associated strain amplitudes may (*i*) grow with time indicating *cyclic hardening under constant stress*, or (*ii*) decay with time indicating *cyclic softening under constant stress*. These features are illustrated in Fig. 6.7. Fig. 6.7(*a*) is an illustration of cyclic hardening under constant stress cycling and Fig. 6.7(*b*) is an illustration of cyclic softening under constant stress cycling. It can be seen from these figures that the stress-strain cycle loops become steeper with increasing time for *cyclic hardening* and the loops become flatter with increase in time for *cyclic softening*.

The challenge in modelling is to capture the features observed in Figs. 6.6 or 6.7. Many times some insight into the mechanisms responsible for the observed behaviour may be useful in developing the models. For example, it is generally accepted that *cyclic hardening under constant stress* for metals is associated with the increase in dislocation density and occurs in metals that are initially soft and with low initial dislocation density. The reverse phenomenon of *cyclic softening under cyclic stress*, occurs in metals which are initially hard. In such metals, the initial dislocation density is observed to be high, and during cyclic straining the rearrangement and annihilation of dislocation substructures causes the overall dislocation density to decrease, leading to cyclic softening. Both cyclic hardening as well as cyclic softening under constant stress are observed to saturate to a constant value after a number of strain reversals, for both the types of materials and reach a stable cyclic behaviour. The capture of these phenomena and linking them with the knowledge of their micromechanical causes, are the challenges faced in the development of various models for cyclic behaviour of these materials.

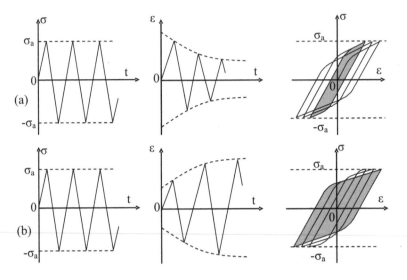

Fig.6.7 Stress and Strain in a Stress controlled test (a) cyclically hardening and (b) cyclically softening materials

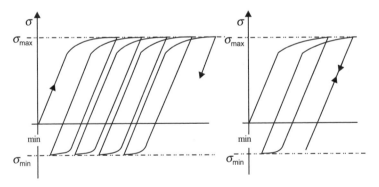

Fig. 6.8 Stress strain response during (a) ratcheting and (b) shakedown

(b) *Ratcheting and shakedown: Ratcheting* and *shakedown* are two phenomena that are likely to be noticed in any material that is subjected to cyclic loading. Both these phenomena relate to the progressive accumulation of plastic strains during cyclic loading and would indicate the desirable characteristics that an elastic-plastic model should exhibit.

Ratcheting is the cycle-by-cycle accumulation of plastic strain for some repetitive loading paths as shown in Fig. 6.8(a). Ratcheting takes place in a material when the material is cyclically loaded at some constant stress/strain amplitude with a non-zero mean stress. During the repetitive loading, each consecutive loop in the stress strain curve, gets displaced forward as shown in Fig. 6.8(a). We observe that during ratcheting, the steady state response of the material is an open elastic plastic loop where the material accumulates a net strain during each cycle.

Shakedown is defined as the condition where by, within a small number of load applications the stress and strain response at all points in the structure reach a steady state of cyclic response as shown in Fig. 6.8(b). As can be seen in the figure, the material that attains a shakedown behavior will show a steady state response is elastic, with no net accumulation of plastic deformation.

Both ratcheting and shakedown are observed in materials even when they are subjected to multiaxial state of loads, where loads along one axis is held constant and the loads along the other axis is varied cyclically. Modelling of this multiaxial behaviour and the prediction of degradation of the material due to cyclic loading, continues to be a challenge for most researchers working on modelling of material response due to cyclic loading.

Cyclic conditions could influence the major decision making in design. Even though each of the above phenomena relates to the deformation response of the material due to cyclic loading, they will also implicitly influence the overall failure of the material due to cyclic loading. The failure of a material due to cyclic loading is called as *fatigue* and this is related to the lowering of the strength of the material due to repeated loading, well below the ultimate strength of the material. It is to be noted that the accumulated plastic strain is a key parameter used in quantifying *fatigue degradation* in most materials.

Example 6.4.1: Consider a material undergoing kinematic hardening. Derive an incremental form of stress-strain relationship for such a material. Following are the yield function, flow rule and hardening rule for the material:

$$\left.\begin{array}{ll} f = |\sigma| - R & \text{yield condition} \\[2mm] de^p = d\lambda \dfrac{\partial f}{\partial \sigma} & \text{flow rule} \\[2mm] dR = EC_R de^p & \text{hardening rule} \end{array}\right\} \qquad \text{(E6.4.1)}$$

Solution:

The yield condition given in Eqn. (E6.4.1) can be translated into a loading condition for the evaluation of f, in the following manner:

$$\left.\begin{array}{l} f = 0 \ \text{condition for yielding} \\[1mm] f < 0 \ \text{condition for elastic loading} \end{array}\right\} \qquad \text{(E6.4.2)}$$

Based on the definition of yield function given in Eqn. (E6.4.1), we can evaluate the partial derivative as

$$\frac{\partial f}{\partial \sigma} = \frac{\partial \left(|\sigma| - R\right)}{\partial \sigma} = sign(\sigma) \qquad \text{(E6.4.3)}$$

where *sign(x)* denotes the sign of value of x. It will be negative for values of x less than zero and will be positive for values of x greater than zero. Based on the flow rule given in Eqn. (E6.4.1), we can write the plastic strain to be

$$de^p = d\lambda \, sign(\sigma) \qquad \text{(E6.4.4)}$$

Since we know the total strain to be sum of elastic and plastic strain (from Eqn. (6.4.12)), we have

$$de^p = de - de^e = de - d\sigma/E \cdot \qquad \text{(E6.4.5)}$$

Rearranging and substituting for de^p, we get

$$d\sigma = E\left[de - d\lambda \, sign(\sigma) \right]. \qquad \text{(E6.4.6)}$$

We know from consistency condition (Eqn. (6.4.11)) that

$$df = 0 \Rightarrow \frac{\partial f}{\partial \sigma} d\sigma + \frac{\partial f}{\partial R} dR = 0. \qquad \text{(E6.4.7)}$$

We note that we are assuming the internal state variable Q defined in Eqn. (E6.4.1), to be the radius R used to define the yield function. Substituting for the partial derivatives and for incremental quantities in the consistency condition Eqn. (E6.4.7), we obtain

$$sign(\sigma) E\left[de - d\lambda \, sign(\sigma) \right] + (-1) E C_R d\lambda = 0. \qquad \text{(E6.4.8)}$$

Rearranging the above equation, we get

$$d\lambda = \frac{sign(\sigma)}{\left\{ sign(\sigma) \right\}^2 + C_R} de. \qquad \text{(E6.4.9)}$$

Substituting Eqn. (E6.4.3) in the expression for the incremental stress in Eqn. (E6.4.6), we obtain (given that $\{sign(\sigma)\}^2 = 1$)

$$d\sigma = E\left[1 - \frac{1}{1 + C_R} \right] de = E\left[\frac{C_R}{1 + C_R} \right] de. \qquad \text{(E6.4.10)}$$

Therefore, we have obtained a relationship between incremental stress and incremental strain of the following form:

$$d\sigma = E\left[\frac{C_R}{1 + C_R} \right] de. \qquad \text{(E6.4.11)}$$

This example demonstrates that the loading parameter $d\lambda$ which normally appears in the definition of the plastic strain, can be eliminated in deriving the relationship between incremental stress and incremental strains. Relationships of the type defined in Eqn. (E6.4.11), are of practical interest, since they will help us to non-linear stress-strain relationship in any material undergoing plastic deformation.

6.4.5 Generalised Description of Plasticity Models

The phenomenon of plastic deformation was described in Section 6.4.1 the modelling of this behaviour is discussed in Sections 6.4.2-6.4.4. For ease of description, the models discussed were primarily one-dimensional in nature. The hyperelastic model discussed in Section 6.2.1 and hypoelastic models described in Section 6.3.1, were introduced as generalized three dimensional models. In order to generalise the procedures outlined in Sections 6.4.2-6.4.4 above, an attempt will be made in this section, to describe these models in a generalised three-dimensional framework and for infinitesimal deformations.

In general, the stress in a plastically deforming material can be given as

$$\sigma = \hat{g}(\mathbf{e}, \mathbf{Q}). \tag{6.4.18}$$

Equation (6.4.18) is a constitutive relation that states that the stress in a plastically deforming material, is a function of a deformation measure $\mathbf{e}$ (strain) and a set of state variables ($\mathbf{Q}$), that are specific to a plastically deforming material. The internal state variables will help in defining the yield stress σ_y.

Following the assumptions made in Eqn. (6.4.3) in decomposing the one dimensional strain into its elastic and plastic parts, it is possible to decompose the three dimensional strain $\mathbf{e}$ also into a plastic three dimensional strain ($\mathbf{e}^P$) and elastic three dimensional strain ($\mathbf{e}^e$) as

$$\mathbf{e} = \mathbf{e}^e + \mathbf{e}^P. \tag{6.4.19}$$

The incremental strains can likewise be decompose into their elastic components and plastic components (analogous to Eqn. (6.4.12)) as

$$d\mathbf{e} = d\mathbf{e}^e + d\mathbf{e}^P. \tag{6.4.20}$$

The elastic component of the strain ($\mathbf{e}^e$), can be related to the total stress σ using the elastic stress-strain relationships developed in Chapter 4. Hence, we can directly use Eqn. (4.2.15), and represent it using tensor notations to be

$$\sigma = 2\mu \mathbf{e}^e + \lambda I_{e^e} \mathbf{e}^e. \tag{6.4.21}$$

The incremental stresses can be written in terms of incremental elastic strains using a constitutive relation of the type

$$d\sigma = \mathbf{C} d\mathbf{e}^e, \tag{6.4.22}$$

where $\mathbf{C}$ is the secant modulus relating the incremental stresses $d\sigma$ and incremental strains $d\mathbf{e}$ (similar to Eqn. (4.2.2)). The estimation of plastic strains has been discussed in detail for incremental loading, in the one dimensional formulation in Eqn. (6.4.14). The total strain is later stated to be a summation of all incremental strains experience by the material up to the current stage of evaluation. The incremental plastic strains were stated to be a function of the yield function f, which in turn was stated to be a function of internal state variable Q.

For a three-dimensional state of stress, the yield function f (Eqn. (6.4.10)) can be redefined to be

$$f = f(\boldsymbol{\sigma}, \mathbf{Q}). \tag{6.4.23}$$

We note that in contrast to Eqn. (6.4.10), $\mathbf{Q}$ used in Eqn. (6.4.23) is a set of variables and will represent a set of state variables that may be associated with plastic deformation. These internal state variables can be in the form of geometric parameters (such as R and C_R) defined in Example 6.4.1, or microstructural parameters such as dislocation density at any material point, or conceptually defined parameters such as back stresses in the material etc.

The criteria for loading or unloading in the material developed in Eqn. (6.4.9), can now be generalized to be defined in terms of gradients of f with respect to the stresses $\boldsymbol{\sigma}$. The generalized criteria, will now be of the form

$$f = 0, \frac{\partial f}{\partial \boldsymbol{\sigma}} \cdot d\boldsymbol{\sigma}(t) > 0 \rightarrow \text{plastic loading},$$

$$f = 0, \frac{\partial f}{\partial \boldsymbol{\sigma}} \cdot d\boldsymbol{\sigma}(t) < 0 \rightarrow \text{elastic loading} \tag{6.4.24}$$

and

$$f = 0, \frac{\partial f}{\partial \boldsymbol{\sigma}} \cdot d\boldsymbol{\sigma}(t) = 0 \rightarrow \text{neutral loading}$$

Further, the flow rule defined in Eqn. (6.4.14), can now be written as

$$d\mathbf{e}^p(t) = d\lambda \frac{\partial f}{\partial \boldsymbol{\sigma}}. \tag{6.4.25}$$

Equation (6.4.25) can be substituted along with Eqn. (6.4.22) into Eqn. (6.4.20) in order to get the following generalized form of incremental relations between incremental stresses and incremental strains (similar to Eqn. (6.4.15)). Hence, we get

$$d\boldsymbol{\sigma} = \mathbf{C}\, d\mathbf{e}^e = \mathbf{C}\left(d\mathbf{e} - d\mathbf{e}^p\right) = \mathbf{C}\left(d\mathbf{e} - d\lambda \frac{\partial f}{\partial \boldsymbol{\sigma}}\right). \tag{6.4.26}$$

The nature of the function f will then determine the exact nature of the relations between $d\boldsymbol{\sigma}$ and $d\mathbf{e}$ in any given material as was illustrated in Example 6.4.1. It must be noted that very often scalar invariants or their equivalent definitions are used in the definition of the function f. This becomes necessary for mathematical manipulations and for physical interpretation of the results.

6.5 CASE STUDY OF CYCLIC DEFORMATION OF SOFT CLAYEY SOILS

Many principles of solid modelling that were outlined in various sections in this chapter, will now be applied to a specific material, *clayey soils*. In this case study, we will examine the general micro-structural features of the material, classical experiments done on the material, the principles of plasticity that are utilized to capture the experimental results through a material model, the estimation of parameters for the model and finally at a comparison of the model predictions with the experimental results.

6.5.1 Material Description

Clayey soils are soils found in coastal regions and are characterised by their ability to resist deformations due to inter-particular friction, as well as inter-particular adhesion due to ionic interaction between particles. Soft clays are saturated clays that posses very low shear strength. Soils have a capacity to hold water in the interstitial space between the grains. Hence, the applied load on the material is visualized to be transferred to water as well as to the soil particles in a soil.

In order to get an idea of the effective load carried by the soil, it is important to define a concept called *pore pressure* in soil mechanics. *Pore pressure* refers to the pressure of groundwater held within a soil (in gaps between particles (pores)). For example, in a soil of high permeability (such as sandy soils), the pore pressure would be equal to the hydrostatic pressure in the soil. However, in a soil of low permeability (such as clayey soils), excess pore pressure will be developed to accommodate the increase in loads on the soil samples. Pore pressure u is often used to define a kinetic variable called *effective stress* σ' which is related to the total stress in the soil σ through the relation

$$\sigma' = \sigma - u\mathbf{I}. \tag{6.5.1}$$

It may be noted here that even though the definition of *effective stress* in Eqn. (6.5.1) looks similar to the definition of *deviatoric stress* defined in Eqn. (3.4.37). However, it must be noted that the concepts are entire different. The pore pressure that is subtracted from the total stress in Eqn. (6.5.1), is a distinct entity and is primarily due the second phase water, that invariably exists along with the soil at any given time. The deviatoric stress defined in Eqn. (3.4.37), is a measure of non-hydrostatic component of any stress in any material and is independent of the existence of a second phase material like water. In soil mechanics, the effective stress σ' is considered to be primary agent that causes displacements in a soil. Hence, constitutive relations are often found between this variable and a kinematic variable such as volumetric strain v in the material.

Modelling the deformational response of soils is very important to predict the stiffness and strength of soils for any load bearing application. Mechanical response of soils is also known to be very sensitive to the pore pressure that will be developed in the material. Pore pressure development is highly sensitive to the type of soil as well as the nature of cyclic loads that may be induced in the soil due to traffic loads or earthquakes. Hence, modelling of the complex mechanical response of the soil is very important for many practical applications.

6.5.2 Experimental Characterization

Cyclic testing in soils are typically done in test setups called as *triaxial test apparatus*. A trial test apparatus along with an indication of location of various measurements is shown in Fig. 6.9. A *triaxial cell* allows a cylindrical sample to be placed in a water chamber, whose pressure can be varied externally. The axial loads in the cylindrical specimen are independently controlled. Hence, the setup allows for an equi-biaxial compression on the sample and an axial variation in loads (either a monotonic or a cyclic load).

During the test, load, deformation and pore water pressure are continuously monitored. The sample is mounted on to the base in the triaxial cell. The load application is through the loading ram fitted on to the cell and the measurements of pore pressure, axial strain etc. are carried out by fixing the necessary gauges in the provisions provided in the cell as indicated in Fig. 6.9.

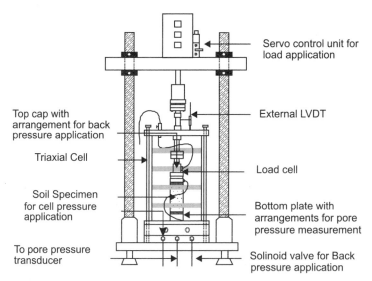

Fig. 6.9 Triaxial cell with measuring devices

The triaxial cell has the facility to conduct the tests on soil samples in the range of 38 mm to 100 mm diameter with confining pressures up to 1000 kPa using the pneumatic control panel. The axial deformation was measured using

LVDT. The pore water pressure was measured with a pore pressure transducer. The axial strains and pore pressures measured in a typical clay sample, are reported in Figs. 6.10 and 6.11, respectively.

6.5.3 Constitutive Model Development for Monotonic and Cyclic Behaviour

It is well known that soils do have a non-linear stress-strain behaviour. In monotonic loading, these non-linear curves were attempted to be captured incremental hypoelastic models. However, all such models will prove to be ineffective, when we have reverse loading in the material. A safe approach that widely accepted, is the cyclic plasticity model, where one identifies a yield function, flow rule, as well the kinetics of the yield function (hardening/ softening) with cyclic loading. A general description of one such plasticity based cyclic model, will be described in this section.

As noted in the description of generalised plasticity models in Section 6.4.5, the following are required to describe the total deformation due to cyclic loading in any material.

(a) A yield function describing the general onset of yield in the material,

(b) A flow rule that helps us to evaluate the incremental plastic strain associated with any loading, and

(c) A hardening law that will help us to describe the evolution of yield surface with number of cycles.

In soil mechanics, *yield* of a soil is termed as a *critical state* (or the failure state) of soil, where the soil experiences large shear deformations at constant volume. As per this theory the state of effective stress of a soil medium at any point of time is expressed in terms of the three main variables, the *effective mean stress p′* and the deviator stress *q*, *p′* and *q* are defined in terms of principal effective stress using Eqn. (6.5.1). Hence, we have

$$p' = \frac{1}{3}\left(\sigma_1' + \sigma_2' + \sigma_3'\right),$$

$$q = \sqrt{\left\{\frac{1}{2}\left[\left(\sigma_1' - \sigma_2'\right)^2 + \left(\sigma_2' - \sigma_3'\right)^2 + \left(\sigma_3' - \sigma_1'\right)^2\right]\right\}}. \qquad (6.5.2)$$

where, σ_1', σ_2' and σ_3' are the principal effective stresses experienced by the soil. In a typical triaxial test, it can be assumed that $\sigma_1' = \sigma_2'$. The loads in the soil are typically monitored through a parameter called as *stress ratio* η which is defined as

$$\eta = q/p'. \qquad (6.5.3)$$

Failure is assumed to take place in the soil when η reaches its critical value η_c. Convenient measures of strain for triaxial conditions are the volumetric strain v, and a shear strain e, which are defined as

$$v = e_1 + 2e_3 \text{ and } e = \frac{2}{3}\left(e_1 - e_3\right). \qquad (6.5.4)$$

where, e_1 and e_3 are the major and minor principal strains respectively in the soil sample.

Using all the stress and strain measures defined in Eqns. (6.5.1) to (6.5.5), it is possible to define the conditions of plastic flow in a soil. A typical flow condition for a clayey soil is listed as:

$$\text{Yield function:} \quad \eta^2 + \eta_c^2 = \frac{\eta_c^2}{\left(\frac{p'}{p_c'}\right)} \quad \text{and} \quad (6.5.5)$$

$$\text{Flow rule:} \quad \frac{\delta v^p}{\delta e^p} = \frac{\eta_c^2 - \eta^2}{2\eta} \quad (6.5.6)$$

where v^p and e^p are the plastic volumetric strain increment and plastic shear strain increment, respectively. It may be noted that Eqn. (6.5.5) is a special form of Eqn. (6.4.10), and Eqn. (6.5.6) is a special form of Eqn. (6.4.14). One may identify the effective mean stress p' as a special form of the state variable Q and the plastic volumetric strain v^p as a special form of the plastic multiplier λ. Hence, we see the constitutive models postulated for soils are based on the concepts of plasticity that were outlined earlier in Section 6.4.3.

When the soil is yielding, the changes in the size of p'_c of the yield locus are linked with the changes in the effective stress p' and $q = \eta p'$, through the differential form of yield surface equation, which is of the form:

$$\frac{dp'}{p'} + \frac{2\eta d\eta}{\eta_c^2 + \eta^2} - \frac{dp_c'}{p_c'} = 0 \quad (6.5.7)$$

Using Eqn. (6.5.7) and the flow rule, a relation between the incremental stresses and incremental strains can be derived (as outlined in Section 6.4.5), so as to obtain a constitutive relation which is given by

$$\begin{pmatrix} dv \\ de \end{pmatrix} = \begin{bmatrix} C_{11} & C_{12} \\ C_{21} & C_{22} \end{bmatrix} \begin{pmatrix} dp' \\ dq \end{pmatrix} \quad (6.5.8)$$

where

$$C_{11} = \frac{(\lambda - \kappa)a}{(1+e)p'} + \left(\frac{\kappa}{(1+e)}\right)\frac{1}{p'}$$

$$C_{12} = C_{21} = \frac{(\lambda - \kappa)}{(1+e)}\left(\frac{1-a}{p'}\right)$$

$$C_{22} = \frac{(\lambda - \kappa)}{(1+e)}\left(\frac{b}{p'}\right) + \left(\frac{1}{3G}\right)$$

$$a = \frac{\eta_c^2 - \eta^2}{\eta_c^4 - \eta^4} \quad \text{and} \quad b = \frac{4\eta^4}{\eta_c^4 - \eta^4}.$$

Where G is the elastic shear modulus of the soil, e is the void ratio in the virgin soil (a soil that has not been subjected to any strain history), and λ and κ are material parameters indicating the state of initial consolidation in the soil. All these parameters can be evaluated by doing appropriate static deformation tests on the soil.

The cyclic behaviour of the soil can be estimated if we supply the *softening rule* (similar to the hardening rule in cyclic plasticity) associated with the response of the soil under cyclic loading. For the current model, the softening rule, indicating the decay of the yield stress p_c' with the number of cycles N, is given using the relation

$$p_c' = \frac{p_i'(N^{-m})(\eta_c^2 + \eta^2)}{\eta_c^2},\qquad (6.5.9)$$

where the initial value (before the application of cyclic loads) of the effective yield stress p_i', and the *degradation parameter m* are obtained from the cyclic response curves of the clayey soil, obtained experimentally. It may be noted that the hardening law used in Example 6.4.1, was in an incremental form. However, the softening law used for clayey soils in the current case study, is in the form of a total yield stress and not in terms of increments of yield stress.

The degradation of the yield stress obtained from Eqn. (6.5.9), is used as an input to estimate the incremental effective stresses dp' and dq at the beginning of each computation. They can also be used to compute the increase in the pore pressure u at the end of each cycle by using Eqn. (6.5.1). Equation (6.5.9) can then be used to obtain the deformations in the soil sample at any given stage of loading.

6.5.4 Comparison of Model Predictions with Experimental Results

Cyclic trial tests conducted on a typical clayey soil had the model parameters as shown in Table 6.3.

Table 6.3 Model Parameters for the Clayey Soil Chosen for the Case Study

S. No.	Parameter	Values	Source
1.	λ	0.17	Consolidation tests
2.	κ	0.016	
3.	η_c	1.2	Effective stress state at failure in static triaxial test
4.	G	10000-15000 kPa	Static triaxial test
6.	m	0.054	Cyclic triaxial test

The five model parameters along with the decay constant obtained from cyclic loading are used and some predictions on the strain development and pore pressure development are made using the constitutive model proposed in this study. The pore pressure and the axial strain development in the clayey soil due to cyclic loads are shown in Figs. 6.10 and 6.11, respectively. The figures show experimental observations from the triaxial test apparatus, as well as the predictions from the constitutive models proposed in this study. The change in response due to different confinement in the model is captured by changing the parameter p_i', in the model. The model predictions capture the general trend of a stiffer response due to additional confinement. There is a greater mismatch between the model predictions and experiments at lower number of cycles, than at higher cycles.

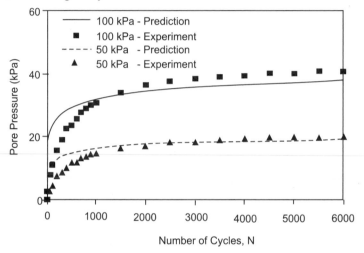

Fig. 6.10 Prediction for pore pressure developed

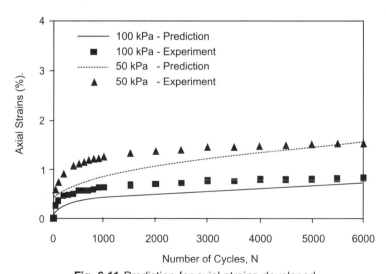

Fig. 6.11 Prediction for axial strains developed

The effects of rate of loading and frequency of loading on the soil response can be incorporated in the current model, by studying their influence on some of the model parameters such as m (degradation parameter).

SUMMARY

In this chapter, we examined three classes of non-linear models that are available to characterize solid deformations. We saw that the non-linear models could be either elastic or inelastic in nature. The non-linear elastic models could be obtained as derivatives of a postulated strain energy density function, and such models are called as hyperelastic models. Under hyperelastic models, we examined the popular models that exist in literature for large deformation of rubbery (elastomeric) materials. The non-linear elastic models could also be obtained as a series expansion of stresses in terms of strains. Such models are called as Cauchy models.

In Section 6.2, we examined the incremental elastic and linear models that can be used to characterize materials that have an overall dissipation (and hence a history dependence), but have a semblance of linearity when analyzed in incremental loading. These models are called hypoelastic models. In Section 6.4, we briefly reviewed the fundamental principles of classical plasticity. We notice that plastic deformation is another mechanism by which a dissipative non-linearity is introduced into the material. We noted that modelling of non-linearity using the theory of plasticity, requires definition of yield condition and flow rules in incremental forms, so that it is possible to trace the non-linear deformations in an incremental fashion. We developed the concepts of plasticity firstly in a one dimensional framework and later have extended the formulation to include the three dimensional state of stress in a material.

In Section 6.5, we have examined a case study of constitutive model development of cyclic response of clayey soils. The constitutive model proposed in this study uses the general theory of plasticity to predict the material response. The use of the proposed model to obtain the non-linear response incrementally during static loading, and the evolution of this non-linear curve due to cyclic load was demonstrated. Typical cyclic tests done on clayey soils were described and the model predictions were compared with the experimental results.

We recognize that non-linear material deformation is quite a complex process at a micromechanical level. For ductile materials, it will be associated with processes like dislocation movements. For brittle materials, it will be associated with formation and development of micro-cracks. For polymers, it will be associated with slippage of molecules. The incorporation of these micro mechanisms is done through the state variables that are defined in the phenomenological models. We have indicated the ways this can be done, while discussing plasticity models in this chapter. Modelling of coupled fields which include mechanical deformation along with other forms of energy, such as thermal fields and electrical fields, will be discussed in Chapter 7.

EXERCISE

6.1 Show that Eqns. (6.2.4) and (6.2.5) reduce to Eqn (6.2.6) for small deformations.

Hint: Use approximations $\mathbf{E} \approx \mathbf{e}, \mathbf{U} \approx \mathbf{I}, \mathbf{F} \approx \mathbf{I}$.

6.2 One of the common applications of strain energy is along with second Piola Kirchhoff stress. Based on definitions given in Table 3.3 and Eqn. (6.2.4), show that

$$\mathbf{S}_1 = \frac{\partial \hat{W}(\mathbf{E}, \theta)}{\partial \mathbf{E}}.$$

6.3 Derive the stress strain relations for a Mooney Rivlin material in simple shear.

6.4 The strain energy density function for a biological tissue such as heart of a bird is given by

$$W = 0.5(e^{1.1(I_C - 3)} - 1).$$

Obtain the stress-strain relations in uniaxial tension for this material.

6.5 Develop the relationships between the coefficients α_1 and α_2 of Eqn. (6.2.31) and show that they can be related to the Lame's coefficients λ and μ that have been defined for isotropic elastic models.

6.6 The total deformation gradient $\mathbf{F}$ can be multiplicatively decomposed into an elastic deformation gradient $\mathbf{F}^e$ and a plastic deformation gradient $\mathbf{F}^p$ as shown below:

$$\mathbf{F} = \mathbf{F}^e \mathbf{F}^p.$$

Show that this multiplicative decomposition of deformation gradients is equivalent to the additive decomposition of strains indicated in Eqn. (6.4.3).

6.7 A long thin-walled cylindrical tank of length L just fits between two end walls when there is no pressure in the tank as shown in figure below. The left wall is fixed in space and the right wall is right wall is flexible enough to accommodate the strains in the tank. The tank is now subjected to an internal pressure p and the material of which the tank is follows a hardening law of the type

$$\sigma_e = \sigma_o + m\left(e^p\right)^n,$$

where e^p is the plastic strain in the material, σ_o, m and n are model parameters and σ_e is the effective stress in the material which is given in terms of the principal stresses σ_1, σ_2 and σ_3 by the expression

$$\sigma_e = \sqrt{\frac{\left(\sigma_1 - \sigma_2\right)^2 + \left(\sigma_2 - \sigma_3\right)^2 + \left(\sigma_3 - \sigma_1\right)^2}{2}} \; .$$

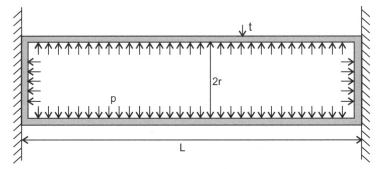

Derive an expression for the internal pressure p when the plastic strain e^p reaches a limiting $\bar{e}$ value in the material.

■■■

7 Coupled Field Response of Special Materials

इयं विद्युत्सर्वेषां भूतानां मध्वस्यै विद्युतः सर्वाणि भूतानि मधु ... *(बृहदारण्यकोपनिषत्)*

iyaṃ vidyutsarveṣāṃ bhootānāṃ madhvasyai vidyutaḥ sarvāṇi bhootāni madhu ... (Bṛhadāraṇyakopaniṣad)

This electricity is the product of all beings and these beings are products of electricity ….

7.1 INTRODUCTION

This chapter presents a brief description of the nature of internal energy changes that are likely to be caused in some materials due to the simultaneous presence of mechanical and non-mechanical fields. We have already seen in Chapter 3 that mechanical deformations may be associated with changes in internal state leading to generation of loss of heat in most materials. Hence, thermodynamics was included as a part of the constitutive formulation in that chapter. It is observed that thermo mechanical interactions are much more pronounced in some special materials like *Shape Memory Materials*. Hence, the modelling of thermo-mechanical response of such materials is developed separately in this chapter. Similarly, mechanical deformations do cause special internal changes, leading to the generation of electrical fields in some special materials like *ferro-electric* materials. The modelling of electro-mechanical response of such materials is also examined closely in this chapter.

Other non-mechanical fields such as magnetic fields are also normally closely associated with electrical fields. Magento-mechanical interactions are specially pronounced in some special materials called *magneto-rheological fluids*, which have a great potential to be used as dampers in many applications. Chemical fields (pH, salt concentration etc.) are also responsible for causing a mechanical response in some other materials. Chemo-mechanical interactions are found to exist in a special class of materials called *stimuli-response gels*. Even though the magneto-mechanical and chemo-mechanical response are not described in detail in this chapter, it is suggested that modelling of such responses are likely to be procedures that are similar to the procedures that are outlined for electro-mechanical and thermo-mechanical response of materials, that are described in some detail in this chapter.

7.1.1 Field Variables Associated with Coupled Field Interactions

The coupled interactions between the electrical, mechanical and thermal fields can be illustrated schematically as shown in Fig. 7.1. This shows two concentric triangles. Each vertex of the triangle represents an energy field. The inner triangle represents the *essential variables* that are associated with the energy field. An *essential variable* is the primary variable that describes the behaviour of any particular field. For mechanical fields, essential variables are the displacements **u**. For thermal fields, an essential variable is temperature θ. Similarly, for electrical fields the essential variable is the electrical polarization **P**. The spatial derivatives of these essential fields such as strains $\left(e = \dfrac{\partial \mathbf{u}}{\partial \mathbf{x}}\right)$ and thermal gradients $\left(\dfrac{\partial \theta}{\partial \mathbf{x}}\right)$ are also treated as essential variables.

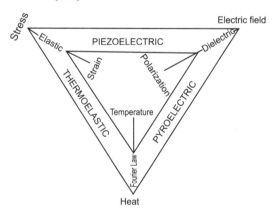

Fig. 7.1 Schematic representation of the coupled interactions between me-chanical, thermal and electrical fields

The three corners of the outside triangle represent *natural variables*, or variables which are necessary to describe the interaction of the material with the external environment. These interactions are normally mathematically described as special balance laws that are specific to the individual energy field. For a mechanical energy field, the natural variable is a field quantity like stress σ. For thermal fields, a natural variable is the heat flux **q** and for electrical fields, a natural variable is *electric field* **E**.

A line joining the two vertices of the inner and the outer triangle at each corner, represents a constitutive relation within each field. For mechanical fields, the constitutive relations are manifested as models of elasticity, viscoelastic and plasticity as discussed in Chapters 4, 5 and 6. For thermal systems the interactions are described as thermal constitutive relations like *Fourier's law*. For electrical systems the interactions are described as electrical relations such as *dielectric equations*. The lines joining the vertices represent coupled interactions. We notice that the coupling could be between the (*a*) essential variables of two fields, or (*b*) natural variables of the two fields or (*c*) Essential variable of one field and the natural variable of another field.

In this chapter, we will examine the electro-mechanical and thermo-mechanical interactions closely, with a view to develop the constitutive relations associated with each of these interactions. In each of these interactions, the basic terms associated with the non-mechanical fields (electrical or thermal), will be briefly reviewed, the balance laws associated with the problems will be stated and the need for constitutive relations for the coupled interactions will be established. Finally, in each of these sections, an example of a constitutive model associated with each coupled interaction will be examined.

7.2 ELECTROMECHANICAL FIELDS

The study of materials which show pronounced electromechanical interactions, has become very important in recent times, because of their use in special devices such as sensors or actuators. For example, piezoelectric ceramic crystals have a great potential to be used as actuators in controls and piezo-polymers have a great potential to be used as sensors in many load cells, hydrophones etc. These materials have the special capability to convert mechanical energy into electrical energy and *vice-versa*. They have a great capacity to store electric charges and are also commonly regarded as good insulators that can sustain large electric fields. Considering the special behaviour of these materials, it is always beneficial to examine the balance laws and the constitutive relations associated with the electromechanical behaviour of these materials. To help in this exercise, the basic terms associated with electrical fields are reviewed in Section 7.2.1.

7.2.1 Basic Definitions of Variables Associated with Electric Fields

Electricity is the energy field associated with the existence of a special scalar property called *electrical charge Q*, in any material. The simultaneous presence of two charges in a material will induce a force of attraction (or repulsion) between these charges. This force normalized per unit charge is called as an *electric field* $\mathbf{E}$. Charges move between two points in space when there is a difference in electrical potential ϕ associated with those points. The flow of charges, denoted as the total quantity of charge moving per unit time across a given cross section is denoted as electric current $\mathbf{C}^{\mathrm{E}}$. Apart from the external field $\mathbf{E}$, process of *polarization* takes place within the material. Denoting $\mathbf{P}$ as the polarized charge per unit area due to polarization, it is possible to write the following expression

$$\mathbf{D}^{E} = \varepsilon_0 \mathbf{E} + \mathbf{P}, \qquad (7.2.1)$$

where ε_0 is the vacuum permittivity. The electric diplacement $\mathbf{D}^{E}$ is a vector field describing the displacement effects of electric field of charges within materials. This quantity is used in Maxwell's equations, as will be described in Section 7.2.2 below.

7.2.2 Balance Laws in Electricity - Maxwell's Equations

The physical principles that relate the various terms that were described in Section 7.2.1, which may also be considered as balance laws in electricity, are popularly termed as *Maxwell's equations* in electrodynamics. There are four Maxwell's equations which are the (a) Gauss law of electricity (which is a consequence of Coloumb's law of electricity), (b) Gauss law of magnetism (law of no magnetic monopoles), (c) Faraday's law (Electric generator principle) and (d) Ampere's law (Current as a source of magnetic field).

Of the four laws listed above, only Gauss law and Faraday's law will be operational in pure electrical fields. These two principles will also be operational when we consider electromechanical responses. Hence, the statements and equations associated with these two principles are reviewed briefly here .These equations will be considered along with the governing equations for mechanical response outlined in Chapter 3, in order to arrive at the mathematical forms that will determine the constitutive relations for electromechanical response in these materials.

Gauss' Law relates the electric field $\mathbf{E}$ flowing across a closed surface, to the electric charge stored within the surface. In the absence of magnetic field and free charges, the integral and the corresponding differential form of Gauss' law reduce to

$$\int_{S^t} \mathbf{n} \cdot \mathbf{D}^{E} \, dA_x = 0 \quad \text{and hence,} \qquad (7.2.2(a))$$

$$\nabla \cdot \mathbf{D}^{E} = 0. \qquad (7.2.2(b))$$

Faraday's Law relates the electric field E to the changes in electromagnetic field in the system. In the absence of any magnetic field, the integral and the corresponding differential form of Faradays' law is of the form

$$\oint_{C} \mathbf{E} \cdot d\Gamma_x = \mathbf{0} \quad \text{and hence,} \qquad (7.2.3(a))$$

$$\nabla \times E = \mathbf{0}. \qquad (7.2.3(b))$$

Equations (7.2.2) and (7.2.3) represent additional equations that need to be satisfied in any material exhibiting an electromechanical response. Very often Eqn. (7.2.3) is further expanded to express the vector electric field E in terms of a scalar electrical potential ϕ using the relations given below:

$$\mathbf{E} = -\nabla\phi, \ \text{ or } \ E_i = -\frac{\partial\phi}{\partial x_i}. \tag{7.2.4}$$

7.2.3 Modifications to Mechanical Balance Laws in the Presence of Electric Fields

The presence of an electric field in a material will involve the consideration of additional forces and moments within each differential volume of a piezoelectric material. We will denote these additional body forces as $\mathbf{b}^{em}$ and the additional body moments (moments per unit volume) as $\mathbf{m}^{em}$. These forces and moments can be obtained from the electric fields using the expressions given below:

$$\mathbf{b}^{em} = (\nabla \cdot \mathbf{P})\mathbf{E} \ \text{ and } \ \mathbf{m}^{em} = \mathbf{E} \times \mathbf{P}. \tag{7.2.5}$$

In the presence of the above forces, there will be an additional working rate $\dot{W}^{em}$ done on the system, which is given by

$$\dot{W}^{em} = \mathbf{b}^{em} \cdot \mathbf{v} + \rho\mathbf{E} \cdot \left(\frac{d\left(\mathbf{P}/\rho\right)}{dt}\right). \tag{7.2.6}$$

Equations (7.2.5) and (7.2.6) may be treated as additional terms to be included in balance laws that have been developed in Chapter 3. In the presence of these forces, moments and energy terms, the integral statements of the balance laws will be take the forms given below:

$$\int_{D^t} \rho(\mathbf{b} + \mathbf{b}^{em})dV_x + \int_{S^t} \mathbf{t}\,dA_x = \frac{d}{dt}\int_{D^t} \rho\mathbf{v}\,dV_x$$

$$\qquad\qquad\qquad\qquad - \text{ Balance of Linear Momentum,}$$

$$\int_{D^t} \left[\left(\mathbf{x} \times \rho\left(\mathbf{b} + \mathbf{b}^{em}\right)\right) + \mathbf{m}^{em}\right]dV_x + \int_{S^t}(\mathbf{x} \times \mathbf{t})dA_x = \frac{d}{dt}\int_{D^t}(\mathbf{x} \times \rho\mathbf{v})\,dV_x$$

$$\qquad\qquad\qquad\qquad - \text{ Balance of Angular Momentum,}$$

$$\int_{D^t} \rho\left(r + \dot{W}^{em}\right)dV_x + \int_{S^t} h\,dA_x + \int_{D^t} \boldsymbol{\sigma}:\mathbf{D}dV_x = \frac{d}{dt}\int_{D^t} \rho\varepsilon\,dV_x$$

$$\qquad\qquad\qquad\qquad - \text{ Energy balance and}$$

$$\int_{D_t} \frac{h}{\theta}dA_x + \int_{D_t} \rho\frac{\left(r + \dot{W}^{em}\right)}{\theta}dV_x \le \frac{d}{dt}\int_{D_t} \rho\eta\,dV_x$$

$$\qquad\qquad\qquad\qquad - \text{ 2}^{\text{nd}} \text{ Law of Thermodynamics.} \tag{7.2.7}$$

We note that the integral statement of the balance of mass will remain unaltered in the presence of the additional body forces and moments. The differential forms of all the above equations will reduce to the following forms:

$$\dot{\rho} + \rho \, \text{div} \, \mathbf{v} = 0$$

$$\text{div } \boldsymbol{\sigma}^T + \rho\left(\mathbf{b} + \mathbf{b}^{em}\right) = \rho a,$$

$$\boldsymbol{\sigma} = \boldsymbol{\sigma}^T,$$

$$\rho\left(r + W^{em}\right) + \text{div}\,\mathbf{q} + \boldsymbol{\sigma}:\mathbf{D} = \rho\dot{\varepsilon}$$

and $\quad \text{div}\left(\dfrac{\mathbf{q}}{\theta}\right) + \rho\dfrac{\left(r + W^{em}\right)}{\theta} \leq \rho\dot{\eta}.$ $\hspace{3cm}$ (7.2.8)

We note that Eqn. (7.2.8) is similar to Eqn. (2.3.52) and represent the mathematical statements associated with each of the physical principles that were discussed before. These equations will have to be solved along with the appropriate boundary conditions, in the presence of additional constitutive relations, which will be discussed in the next section.

We note that the total number of unknowns in Eqn. (7.2.7) are (*a*) density, (*b*) three components of velocities, (*c*) six components of Cauchy stresses, (*d*) three components of heat flux, (*e*) internal energy, (*f*) temperature, (*g*) entropy, (*h*) three components of electric field (*i*) three components of electrical displacements and (*j*) one component of free charge density. Therefore, the total number of unknowns reduces to 23.

Observing Eqn. (7.2.8), we observe that we have (*a*) one equation for continuity of mass, (*b*) one equation for energy, (*c*) three equations for linear momentum balance, (*d*) one Gauss law equation and (*e*) three Faraday's law equations. Hence, we have in total nine balance equations. Thus, we require additional relations that would make the system complete and determinate. These additional relations are supplied by constitutive relations, which will be similar to the constitutive relations that were outlined in Chapter 3. These constitutive relations will be explained in detail in the next section.

7.2.4 General Constitutive Relations Associated with Electromechanical Fields

Following the procedure of postulating constitutive relations for thermoelastic materials, it is possible to generate constitutive relationships for a piezoelectric material that will describe relations for stress $\boldsymbol{\sigma}$, internal energy ε and the entropy η. The independent variables that are to be identified for each of these quantities would be a kinematic variable such as $\mathbf{C}$ associated with motion, as well as electrical variables such as electrical field $\mathbf{E}$. Hence, the constitutive relations for a piezoelectric material can be postulated to be:

$$\psi = \hat{\psi}\left(\mathbf{C}, \theta, \mathbf{E}\right),$$

$$\hat{\mathbf{S}} = 2\rho_o\mathbf{F}\frac{\partial\hat{\psi}\left(\mathbf{C}, \theta, \mathbf{E}\right)}{\partial\mathbf{C}},$$

$$\hat{\eta} = -\frac{\partial \hat{\psi}(\mathbf{C}, \theta, \mathbf{E})}{\partial \theta} \quad \text{and}$$

$$\mathbf{D}^E = \hat{\mathbf{D}}^E(\mathbf{C}, \theta, \mathbf{E}). \tag{7.2.9}$$

We note that Eqn. (7.2.9) is similar to Eqn. (3.4.27) introduced in Chapter 3. They differ from Eqn. (3.4.27) in the following two aspects:

(*a*) Presence of an additional variable $\mathbf{E}$ in the general definition of ψ.

(*b*) Presence of an additional constitutive relation for the electric displacement $\mathbf{D}^E$.

Let us focus our attention on electromechanical behaviour under isothermal conditions. For these conditions, we could remove temperature θ as an independent variable. Further, for materials undergoing infinitismal deformations, the right Cauchy Green tensor $\mathbf{C}$ can be replaced with infinitesimal strain measure $\mathbf{e}$, as was done earlier in Eqn. (6.2.6) in Chapter 6. Additionally, since there is very little distinction between Cauchy stress σ and the Piola-Kirchhoff stress $\mathbf{S}$ for small deformations, we could replace $\mathbf{S}$ with σ in Eqn. (7.2.9). We also state here (without proof) that it is possible to derive an expression for $\mathbf{D}^E$ as a gradient of potential ψ with respect to the electric field $\mathbf{E}$. Incorporating all these modifications, the general form of the constitutive relations given in Eqn. (7.2.9), would become

$$\psi = \hat{\psi}(\mathbf{e}, \mathbf{E}),$$

$$\sigma = \hat{\sigma}(\mathbf{e}, \mathbf{E}) = \frac{\partial \hat{\psi}(\mathbf{e}, \mathbf{E})}{\partial \mathbf{e}} \quad \text{and}$$

$$\mathbf{D}^E = \hat{\mathbf{D}}^E(\mathbf{e}, \mathbf{E}) = -\frac{\partial \hat{\psi}(\mathbf{e}, \mathbf{E})}{\partial \mathbf{E}}. \tag{7.2.10}$$

Equation (7.2.10) gives the desired mathematical form for the constitutive relations associated with small deformations in an electro-mechanical material. In Section 7.2.5, we will expand on the specific forms of *linear* electromechanical constitutive relations.

7.2.5 Linear Constitutive Relations Associated with Electromechanical Fields

Let us assume a simple form of the scalar potential ψ to be of the form

$$\psi = \mathbf{Cee} - \kappa \mathbf{EE} - \varepsilon \mathbf{Ee} \tag{7.2.11}$$

Substituting Eqn. (7.2.11) in Eqn. (7.2.10) and using Eqn. (6.2.6), it is possible to derive the following relations for σ and $\mathbf{D}^E$:

$$\sigma = \mathbf{Ce} - \varepsilon \mathbf{E} \quad \text{and}$$

$$\mathbf{D}^E = \kappa \mathbf{E} + \varepsilon \mathbf{e}. \tag{7.2.12(a)}$$

Equation (7.2.12(a)) can be written in its component form as

$$\sigma_i = C_{ij}e_j - \varepsilon_{ij}E_j \text{ and}$$

$$D_i^E = \kappa_{ij}E_j + \varepsilon_{ij}e_j. \qquad (7.2.12(b))$$

While writing the index representation in Eqn. (7.2.12(b)), we are following the single index notation for stresses and strains, that were introduced in Section (4.2.2) in Chapter 4. It may be noted that in writing the expressions in Eqn. (4.2.12(b)), the single index representation was used for the stresses σ and strains e. It may also be noted that Eqn. (7.2.12) accounts for the possible anisotropy in the material constants κ and ε. We observe that κ is a 3×3 matrix having 9 coefficients and ε, which is normally called as *piezoelectric coefficient* matrix having a size of 3 × 6 with 18 coefficients.

We note that the coefficient matrix C_{ij} appearing in Eqn. (7.2.12) is the same as the coefficient appearing in Eqn. (4.2.1) associated with the linearized stress-strain relations outlined in Chapter 4. We further note that the coefficient ε appearing before **E** in Eqn. (7.2.12), is the same coefficient that appeared before **e** in Eqn. (7.2.11). This equivalence of the coefficients comes from a rigorous mathematical proof considering the thermodynamic equation of state of the material, and is accepted to be true in the current description. The negative sign associated with the coefficient ε comes in front of the second term because, a compressive strain due to applied voltage on being restrained, induces a tensile stress in the material and *vice versa*.

7.2.6 Biased Piezoelectric (Tiersten's) Model

The models that are discussed in Sections 7.2.4 and 7.2.5 are based on total deformations or total electrical fields and are analogous to the nonlinear hyper-elastic models that were described in Section 6.2.2. Such models, as we have seen earlier, are not sensitive to incremental changes or rates of change in various kinematic and kinetic variables. As we have seen earlier in Section 6.3.1, the incremental models such as the hypoelastic models are more amenable to model the effects of incremental or rates of loading. Most piezoelectric materials are known to respond primarily to incremental changes of loading or voltages and not necessary to total strain or voltage available in the material. These incremental loadings (or voltages) are often applied over a bias strain (or voltage). Hence, there is a need to develop models that can capture the response of the material to incremental loads applied with a certain amplitude and frequency, over a bias strain. Such models are called as biased models and are described in some detail in this section.

We recall that the incremental models that were developed under hypoelastic models of Section 6.3.1. A linearized constitutive relation for mechanical deformation (written in tensor notation) was derived to be

$$\dot{\sigma} = \mathbf{C}(\sigma)\dot{\mathbf{e}}, \qquad (7.2.13)$$

where we recognize that $\dot{\sigma}$ and $\dot{e}$ denote incremental changes $d\sigma$ and de for deformations that are homogeneous in time. We could develop incremental models for piezoelectric material deformation also on similar lines by linearizing the general dependence of σ and $\mathbf{D}^E$ on the kinematic variables $\mathbf{e}$ and $\mathbf{E}$ as shown in Eqn. (7.2.10). Hence, it is possible to write a general incremental constitutive relation for a piezoelectric material as

$$\dot{\sigma} = \mathbf{C}(\sigma)\dot{e} - \varepsilon(\mathbf{E})\dot{\mathbf{E}} \text{ and}$$

$$\dot{\mathbf{D}}^E = \kappa(\mathbf{E})\dot{\mathbf{E}} + \varepsilon(\mathbf{E})\dot{e}. \qquad (7.2.14)$$

where we note that the constants $\mathbf{C}(\sigma)$, $\varepsilon(\mathbf{E})$ and $\kappa(\mathbf{E})$ are dependent on the variables σ and $\mathbf{E}$.

Assume that a piezoelectric material is being subjected to a loading over a biased electric field $\mathbf{E}^0$ and a biased strain of $\mathbf{e}^0$. For materials operating under such conditions, a linearized electro-elastic model with small dynamic fields (electrical and mechanical fields) superimposed on a static bias (electrical and mechanical fields), could be developed and written as

$$\dot{\sigma} = \mathbf{C}^0\dot{e} - \varepsilon^0\dot{\mathbf{E}} \text{ and}$$

$$\dot{\mathbf{D}}^E = \kappa^0\dot{\mathbf{E}} + \varepsilon^0\dot{e}. \qquad (7.2.15)$$

where $\mathbf{C}^0$, ε^0 and κ^0 are the elastic modulii, piezo constant and the dielectric constant about the biased state.

Equation (7.2.15) is useful from the practical point of view for poly vinylidene fluoride, PVDF (also discussed in Section 1.2) sheets which are generally subjected to prestress in applications in which they are used. For example, in hydrophones and stethoscopes, PVDF used is subjected to a pre-stress and a mechanical pulse is applied on these pre-stretched sheets. The response of these sheets can be obtained effectively using the model described in Eqn. (7.2.15).

7.3 THERMOMECHANICAL FIELDS

Thermomechanical effects are normally observed in most materials. For example, it is a common observation that when we supply heat to any material, the associated changes in temperature will also produce strains, called as *thermal strains*, depending upon the boundary conditions of the problem. Similarly, mechanical loads applied at very high rates, may result in internal dissipation of mechanical energy, leading to heat generation and an increase in temperatures in any material. However, the conversions from mechanical energy to thermal energy, is insignificant in most materials. In some special materials, these conversions are much more pronounced, primarily because of the internal phase transitions that take place in these materials. The significant thermomechanical conversions that are seen in these materials make them attractive candidates for use as thermal sensors or as thermally induced phenomena. In this section, we will discuss the constitutive relations associated with one such class of material called as the *shape memory materials*.

The term *shape memory* is associated with these materials, because these materials have a tendency to remember the residual strains that they were subjected to, and to supply back those strains when they are heated. Shape memory behaviour associated with metals are primarily associated with alloys such as Nickel-Titanium Alloys and these materials are called as *shape memory alloys* (SMA). Polymers that exhibit shape memory effect are called as *shape memory polymers* and many compounds of polyethylenes and polyurethanes are known to exhibit shape memory effect. Shape memory materials have received increasing attention in the recent years, because of their characteristic large recoverable strains and high actuation force (when subjected to thermal loading), where the speed of response is not an important consideration in design.

7.3.1 Response of Shape Memory Materials

Shape memory materials are known to have distinctly two special properties. These are (*a*) *Shape memory effect* and (*b*) *Finite Elasticity*. Any plastically deforming material will develop some residual strains when it is subjected to a cycle of loading and unloading. The recovery of these residual strains due to heating, is called as the *shape memory effect*. The capacity of any material to undergo finite (large) strains on loading is the *finite elastic effect* and this behavior is found in all shape memory materials. Each of these behaviours will be illustrated separately with reference to metals and polymers in the following subsections.

7.3.1.1 Response of Shape Memory Alloys

Fig. 7.2 shows a typical thermomechanical response of a shape memory alloy. The response of the alloy needs to be noted at three distinct temperatures, which are denoted as the low temperature θ_l, high temperature θ_h and very high temperature θ_{vh}. At θ_l the response of the alloy is like any other plastically deforming metal. In other words, the material undergoes yield causing sufficiently large plastic deformations, hardens further on application of loads and has a residual plastic strain on complete unloading. The response of the material at very high temperatures θ_{vh} is also like that of any other metal at high temperature. At such temperatures, one does not find any significant yield, but the material melts after reaching a high yield stress.

The behaviour of SMA at the high temperatures θ_h is interesting to note. The material yields on loading, and accumulates a considerable amount of plastic deformation and later hardens taking some additional loads. On unloading, the material does go completely to a stage of zero strain (not observed at θ_l).

We find that the material recovers its plastic strain (at a lower stress than the yield stress) and joins the initial elastic curve. Thereafter, the material retraces the elastic deformation line and goes back to state of zero stress and zero strain. This is the finite elasticity behaviour of shape memory alloy, where we observe that material undergoes large deformations and returns to the original state of stress and strain. In other words, the overall behaviour of the material is like a rubber, undergoing large elastic deformations. However, since the material starts retracing its path at a much lower stress, one can observe a hysteresis loop (indicating some energy dissipation) at the end of one cycle of loading. Hence, the *finite elastic* behaviour of SMA at this temperature is normally termed as *superelasticity* and is an important feature that needs to be modeled if the material is used at such temperatures.

It is further interesting to note what happens to the material if the material after unloading at θ_l is then heated to θ_h. As observed in Fig. 7.2, all the residual strain is completely recovered when the material is heated to θ_h. Further, the material returns back to a zero state of stress and strain upon cooling back to θ_l. We observe that the plastic deformations that induce shape to the material are remembered by the material upon heating and hence this effect is called as the *shape memory effect*. The strains that are recovered during heating release strain energy and hence, this released energy is used in applications that involve actuation.

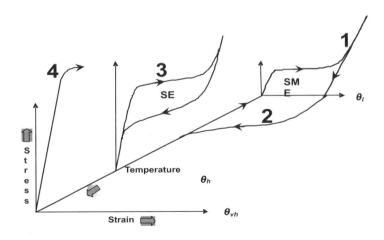

Fig. 7.2 Typical thermomechanical response of a shape memory alloy

7.3.1.2 Response of Shape Memory Polymers

The thermomechanical behaviour of a shape memory polymer is illustrated in Fig. 7.3. It must be noted that the mechanical behaviour of polymers at high temperatures is similar to a nonlinear hyperelastic finite deformational response (of rubber like materials), which was described in Section 6.2.2. Assume that a shape memory polymer is subjected to some mechanical strain at a high temperature θ_h as shown in Fig. 7.3. The material is then cooled to the lower temperature θ_l. We note that unlike metals, this cooling is associated with a change of stresses in the material. When this stress is released at θ_l, we note that there is a residual strain available in the material at θ_l. This residual strain in the material is recovered when the temperature is raised back to θ_h and this step is very similar to the response of alloys that was shown in Fig. 7.2. We note that shape memory polymers differ from shape memory alloys in the *training sequence*, *i.e.*, in the thermo-mechanical cycle that induces residual strains at low temperatures. While the mechanical cycling to accumulate plastic strains is primarily done at θ_l for shape memory alloys, the mechanical cycling to accumulate strain energy is primarily done at θ_h for shape memory polymers. The microstructural changes that are responsible for the macroscopic behaviour is totally different in the two shape memory materials and these changes are described briefly in Section 7.3.2 below.

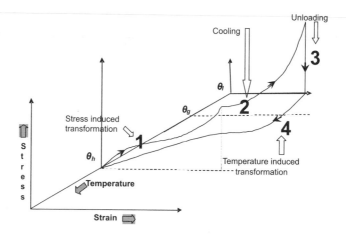

Fig. 7.3 Typical thermomechanical response of a shape memory polymer

7.3.2 Microstructural Changes in Shape Memory Materials

The shape memory effect in shape memory materials is primarily because of certain microstructural changes that occur in the material. A rudimentary knowledge of these microstructures and their transformations will help us to appreciate the material behaviour in a better way. This understanding will also help us to correlate the macro mechanical behaviour of the material (as observed in the stress-strain relations) with the microscopic changes in the material. Hence, the microstructures and the transformations associated with these microstructures, during deformations associated with shape memory effect will be examined briefly below.

7.3.2.1 Microstructural Changes Associated with Shape Memory Alloys

The microstructure of a shape memory alloy is characterized by the existence of the material in a certain crystallographic phase at any given state of stress, strain and temperature. Shape memory alloys are known to exist primarily in two different phases in stress free configurations. One of these phases which is stable at low temperatures is called as the *martensite* (M) phase and the other phase that is stable at high temperatures is called the *austenite* (A) phase. The martensite phase further exists in two structural forms called *twinned martensite* (TM) and *detwinned martensite* (DM). Hence, at any given load and temperataure, the material can exist either as (*a*) *Detwinned martinsite*, (*b*) *Twinned martinsite* or (*c*) *Austenite*.

As we increase the stresses or temperature, transformations take place between (*i*) detwinned martensite to twinned martensite (transformation 1), or (*ii*) Detwinned martensite and austenite (transformation 2) or (*ii*) Austenite to twinned martensite (transformation 3). These transformations are illustrated schematically in the phase diagram shown in Fig. 7.4.

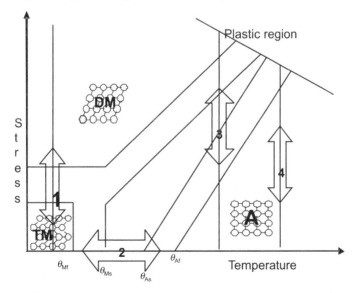

Fig. 7.4 Phase transformations associated with shape memory alloys

It may be noted in Fig. 7.4 that phase transformations associated with shape recovery (plastic strain recovery), caused by increase in temperature, are characterized by two transformation temperatures, *viz.*, austenite start temperature θ_{As} (temperatures at which martensite-austenite transformations begin to take place) and austenite finish temperature θ_{Af} (temperature at which transformations are completed). Similarly cooling of the temperature results in self- accommodation and is associated with two distinct temperatures, *viz.*, martensite start temperature θ_{Ms} (temperatures at which austenite-martensite transformations begin to take place) and martensite finish temperature θ_{Mf} (temperatures when the transformations are complete). These temperatures of the material are highly sensitive to alloy compositions, heat treatment and cold-working.

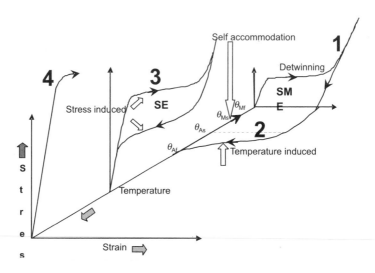

Fig. 7.5 Thermomechanical behaviour of SMA indicating various transformations occurring at various stages.

The transformations that are likely to take place at different stage of loading in a test specimen are indicated in Fig. 7.5, which is a repetition of Fig. 7.2, except that the transformations occurring at various stages are indicated clearly. The identification of the various transformation temperatures and the volume fractions of any material, in a given phase, will go a long way in constructing constitutive models that predict the experimental observations based on the knowledge of the microstructure at critical phases. These constitutive models will be reviewed briefly in Section 7.3.3.

7.3.2.2 Microstructural Changes Associated with Shape Memory Polymers

Shape memory polymers are essentially polymers whose internal structure is amenable for pronounced shape memory effect. It is well known that polymers are long chain molecules with a number of repeating units. Polymers can have a highly disordered phase, *the amorphous phase*, and a ordered phase, *the crystalline phase*. *Glass transition temperature* is observed in the amorphous phase, above which polymer is rubbery and flexible, while below which the polymer is glassy and rigid. Melting transition is associated with the crystalline phase, above which polymer exists in a liquid phase. Based on the microstructural changes during glass transition and melting transition, several different types of shape memory polymers have been developed. In the following discussion, shape memory effect and the microstructural changes based on glass transition are described. The temperatures of interest to describe the shape memory behaviour, are θ_g (glass transition temperature), θ_h (high temperature, training temperature) and θ_l (low temperature).

It may be noted that the response of the polymer at either of the two temperatures, θ_h and θ_l, is primarily elastic. This means that load reversal at either of the temperatures will not cause significant plastic deformations. This is different from the behaviour of shape memory alloys shown in Fig. 7.5, where we note that there is an accumulation of plastic deformation at θ_l. We also note that although there is no residual plastic deformation at θ_h in case of shape memory alloys, every loading cycle is likely to result in dissipation of energy due to a phenomenon called superelasticity. On the other hand, in case of shape memory polymers, very large deformations are possible at θ_h, and these deformations are completely reversible (similar to hyperelastic deformation in elastomers that are discussed in Section 6.2.2).

The mechanisms that are postulated to be responsible for the elastic behaviour at both the temperatures are different. The elasticity at θ_l is the elasticity of the glassy phase and is known to be because of stretching of primary bonds and secondary intermolecular forces. The elasticity of the polymer at θ_h is postulated to be due to *entropy elasticity*. Entropy elasticity is a phenomenon that occurs due to decrease in entropy of the material in the amorphous phase due to external load. The entropy of these long chained molecules is again regained when the polymer is relaxed due to removal of load. This recovery of the original entropy is responsible for elasticity in the polymer at θ_h.

The training of polymer is carried out at θ_h and accumulation of strains takes place when the polymer is cooled in a *stretched or trained* state from θ_h to θ_l. The recovery of accumulated plastic strains, which is the shape memory effect, takes place when the polymer is heated from θ_l to θ_h in a stress free state. We note that the polymer undergoes glass transition between these two temperatures. Hence, knowledge of θ_g and possible microstructural changes that are likely to take place at θ_g are very important in modelling the shape memory effect in polymers.

Glass transition in polymers related to mobility of molecular segments (parts of polymer chain). Above the glass transition, associative motion of polymer segments is possible, while below glass transition the associative motion is "frozen" or not possible. When polymer is cooled from θ_h to θ_l (across glass transition) in a stretched state, strain gets "locked in" due to lack of mobility of molecules as temperature reaches around and lower than θ_g. Once in the glassy state, the segmental motion remains frozen and therefore, the strain remains "locked in". When the polymer is heated from θ_l to θ_h in a stress free state, molecular mobility is possible above glass transition. Since the segmental motion is possible, molecules tend to relax rather than remain in the *locked in* stretched state. These molecular relaxations are responsible for recovery of strain.

7.3.3 Constitutive Modeling of Shape Memory Materials

As described in Section 7.3.2, the thermo-mechanical response of a shape memory material is primarily due to phase transformations that are taking place within the material. The transformations are either in the form of a solid-solid transformation between *martensite* and *austenite* in a shape memory alloy, or in the form of a glass transition in a shape memory polymer. It would be useful to identify the volume fraction of any one of the phases as an additional internal variable which will also influence the thermo-mechanical response of these materials. Other internal variables can also be defined, which capture the microstructure. Hence, example constitutive models are illustrated below, which attempt to capture the mechanical response described in Section 7.3.1, will necessarily have these additional variables. The details of the model will be explained separately for each material below.

7.3.3.1 Constitutive Models for Shape Memory Alloys

In shape memory modelling, it is conventional to denote volume fraction of martensite phase of the material ξ as the independent variable, while developing the constitutive relations. Further, the strains associated with all transformations are small enough, for us to deal with infinitesimal deformation measure $\mathbf{e}$, rather than a finite strain measure $\mathbf{C}$, $\mathbf{B}$ or $\mathbf{E}$. Since all transformations are temperature dependent, it is not possible for us to ignore temperature θ as a variable. Hence, following the logic of the constitutive equations of piezoelectric materials expressed in Eqns. (7.2.9) and (7.2.10) it is possible for us to write following relations for a shape memory alloy:

$$\psi = \hat{\psi}(\mathbf{e}, \xi, \theta)$$

$$\boldsymbol{\sigma} = \hat{\boldsymbol{\sigma}}(\mathbf{e}, \xi, \theta) = \rho_o \frac{\partial \hat{\psi}(\mathbf{e}, \xi, \theta)}{\partial \mathbf{e}} \quad \text{and}$$

$$\eta = -\frac{\partial \hat{\psi}(\mathbf{e}, \xi, \theta)}{\partial \theta}. \tag{7.3.1}$$

An equation for increments in stresses can be easily derived by expanding the expression for stresses in Eqn. (7.3.1) as

$$d\boldsymbol{\sigma} = C(\mathbf{e}, \xi, \theta) d\mathbf{e} + \Omega(\mathbf{e}, \xi, \theta) d\xi + \Theta(\mathbf{e}, \xi, \theta) d\theta$$

where $\quad C(\mathbf{e}, \xi, \theta) = \dfrac{\partial \boldsymbol{\sigma}}{\partial \mathbf{e}}; \Omega(\mathbf{e}, \xi, \theta) = \dfrac{\partial \boldsymbol{\sigma}}{\partial \xi}$ and $\Theta(\mathbf{e}, \xi, \theta) = \dfrac{\partial \boldsymbol{\sigma}}{\partial \theta}. \tag{7.3.2}$

Equations (7.3.1) and (7.3.2) suggest that an assumed form of the Helmholtz free energy $\psi = \hat{\psi}(\mathbf{e}, \xi, \theta)$ will be sufficient, to derive expressions for the tangent modulus $C(\mathbf{e}, \xi, \theta)$ or the transformation modulus $\Omega(\mathbf{e}, \xi, \theta)$ in the material. Alternatively, the evolution of the tangent modulus, as observed from the experimental stress-strain graphs may be postulated and other functions can be derived from the same.

The variation of the internal variable ξ may be chosen to conform to the changes in phase diagrams as shown in Fig. 7.4. Representative values of changes in Nitinol, and their use in tracing the isothermal stress-strain curve of the material, are indicated in Exercise 7.2. It has been observed by researchers that modelling of a variety in material response due to complex loading situations like cyclic loading, inner hysteresis loops etc. is easier handled using an assumed form of energy function. More details of the same may be found in the references indicated in the bibliography.

It may be noted that the detailed tracking of the stress-strain curves is possible in a shape memory alloy, since the mechanism of its transformation, from the standpoint of its microstructure is fairly well defined and understood. Hence, attempts to link the microstructure to the overall thermomechanical response, are successful in this material. Such detailed correlation is difficult in complex behaviour such as cyclic behaviour of clays (illustrated in Section 6.5). We will also observe that such a predictive capability is not attempted in other shape memory materials such as shape memory polymers, as will be illustrated below in Section 7.3.4.

7.3.3.2 Constitutive Models for Shape Memory Polymers

In Section 7.3.3.1, it was mentioned that volume fractions of Austenitic phase ξ, is used as independent variable while developing the constitutive relations. Similarly, a variable indicating the state of microstructure of the polymer is also required for constitutive modeling of shape memory polymers. It may be noted that the transition from θ_h to θ_l (Fig. 7.4) or vice versa is due to change in mobility and flexibility of molecules. It is possible to visualize this change in terms of variation of *frozen – rubber* phases in polymers. Such models treat ξ as the volume fraction of *frozen* phase. Therefore, ξ evolves with temperature. Unlike in case of shape memory alloys, no distinct phases can be identified in case of shape memory polymers. Therefore, in constitutive modelling of shape memory polymers, alternate variables related to microstructural changes can be defined.

Other representations of the dependence of material behaviour on temperature include a *locked-in* strain e_p and temperature dependent *yield stress* σ_p. Therefore, similar to Eqn. (7.3.1), an example starting point for the modelling of shape memory polymers is given by:

$$\psi = \hat{\psi}\left(\mathbf{e}, e_p, \theta\right)$$

$$\sigma = \hat{\sigma}(\mathbf{e}, e_p, \theta) = \rho_o \frac{\partial \hat{\psi}\left(\mathbf{e}, e_p, \theta\right)}{\partial \mathbf{e}} \quad \text{and}$$

$$\hat{\eta} = -\frac{\partial \hat{\psi}\left(\mathbf{e}, e_p, \theta\right)}{\partial \theta}. \qquad (7.3.3)$$

Stresses can be evaluated using equation that is similar to Eqn. (7.3.2). Using Eqn. (7.3.2), shape memory cycle in polymers can also be modelled (Exercise 7.3).

SUMMARY

This chapter examined the mechanical behaviour of certain special materials, whose mechanical behaviour is strongly influenced by non-mechanical fields. The two specific responses that are looked in this chapter, are the electro-mechanical response and the thermo-mechanical response in these special materials. While the electro-mechanical response was examined in ferroelectric materials, the thermo-mechanical response was examined in shape memory materials.

The characterization of electro-mechanical response of ferroelectric materials requires the consideration of additional field equations and additional field variables, while developing their constitutive models. A linearized form of the ferroelectric model that is useful for piezoelectric activation of a piezoceramic was described in this chapter. The electromechanical response of a piezoelectric polymer, that normally operates in the presence of a biased stress, was also examined in this chapter.

Shape memory effect was examined with reference to its presence in shape memory metallic alloys as well as in shape memory polymers. The necessity of an additional internal variable to capture the total (loading as well as unloading) response of these materials was examined. The internal variable was found to be in the form of volume fraction of a particular phase called martensitic phase in a shape memory alloy. Using this internal variable, the mathematical tools that are required to capture the recovery of a residual strain with increase of temperature, were developed in this chapter. The internal variable for a polymer was postulated to be in the form of a concept like frozen fraction, or in the form of a locked-in strain in the polymer.

EXERCISE

7.1 A sample piezoelectric actuator is shown in figure below along with the orientation axis of the material.

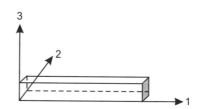

When voltage is applied to this material along its 3-direction, it induces strains in the 1-direction. This strain gets transmitted to a beam structure that is oriented in the 1-direction as shown in figure below.

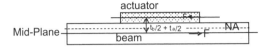

One of the models that can be used to transfer of strains from the actuator to the beam is the *uniform strain model*, which assumes that the strain in the actuator e_a and the strain in the beam e_b are equal and are given by:

$$e_b = e_a = \frac{V d_{31}}{t_a \left(1 + \dfrac{E_b t_b}{E_a t_a} \right)}$$

where E_b is the modulus of elasticity of the beam, E_a is the modulus of elasticity of the actuator (PZT), t_a is the actuator thickness, t_b is the beam thickness, V is the electrical potential applied across the actuator electrodes and d_{31} is the piezoelectric electromechanical coupling coefficient that correlates applied voltage and induced mechanical strain.

Prove that the moment generated by the application of voltage on the actuator is given by the expression:

$$M = F \left(\frac{t_b}{2} + \frac{t_a}{2} \right) = d_{31} \times \left(\frac{t_b + t_a}{2} \right) \times \frac{E_b t_b E_a w_b}{\left(E_b t_b + E_a t_a \right)} \times V .$$

7.2 Consider a Shape Memory Alloy *Nitinol,* with the material properties as shown in Table below.

Moduli	Transformation Temperatures	Transformation Constants	Maximum Residual Strain
$C_s = 67 \times 10^3$ MPa	$\theta_{Mf} = 9\ ^\circ C$	$\sigma_s^{cr} = 100\,\text{MPa}$	$e_L = 0.067$
$C_m = 27.3 \times 10^3$ MPa	$\theta_{Ms} = 18.4\ ^\circ C$	$\sigma_f^{cr} = 170\,\text{MPa}$	
	$\theta_{As} = 34.5\ ^\circ C$	$C_M = 8$ MPa/ $^\circ C$	
	$\theta_{Af} = 49\ ^\circ C$		

Assume that the material follows a Brinson's model which is characterized with the following relations in uniaxial loading:

$$\sigma = C(\xi)(e - e_L \xi_S)$$

$$\xi_s = \frac{1}{2} \cos \left\{ \frac{\pi}{\sigma_S^{cr} - \sigma_f^{cr}} \left(\sigma - \sigma_f^{cr} - C_M (\theta - M_s) \right) \right\} + \frac{1}{2}$$

$$\xi_T = 0,$$

$$\xi = \xi_S + \xi_T \text{ and}$$

$$C(\xi) = C_a + \xi(C_m - C_a)$$

Plot the isothermal stress-strain curve of Nitinol at a temperature $\theta = 20\ ^\circ C$. Plot the variation of the curve for stress ranging from 0-200 MPa. Plot the stress-strain curve for a full cycle of loading and unloading of the specimen.

7.3 Consider a shape memory polymer being governed by the following expression for Helmholtz free energy (one dimensional deformation):

$$\psi(\theta, e, e_p) = \frac{1}{2} B(e - e_\theta)^2 + \frac{1}{2} A(e - e_\theta - e_p)^2 + \psi_\theta$$

where θ = temperature of the material in Kelvin

      e = Total strain

      e_p = Locked up plastic strain

      e_θ = Thermal strain

      B = $B(\theta)$ denotes the entropic elasticity stiffness

      A = $A(\theta)$ denotes the glassy phase stiffness

      ψ_θ = $\psi_\theta(\theta)$ denotes thermal contribution towards ψ.

Derive the expression for stress as a function of strain.

■ ■ ■

8

Concluding Remarks

पूर्णमदः पूर्णमिदं पूर्णात् पूर्णमुदच्यते । पूर्णस्य पूर्णमादाय पूर्णमेवावशिष्यते ।।
(बृहदारण्यकोपनिषत्)

Pūrṇamadaḥ pūrṇamidaṃ pūrṇāt pūrṇamudacyate ।
Pūrṇasya pūrṇamādāya pūrṇamevāvasisyate ।।
(Bṛhadāranyakopaniṣad)

That (primordial cause) is complete. This (effect in the form of universe) is complete. From that completeness, this completeness is born. By knowing the completeness of this (effect) complete entity, completeness (of the cause) alone remains.

8.1 INTRODUCTION

This book has attempted to be a treatise of material models that are normally used in engineering practice. Many material models that are used for solids and fluids were reviewed in this book. The framework of continuum mechanics was used to describe the material models, and it was noted that the material models are known as *constitutive models* in continuum mechanics. It was noted that constitutive models were visualized as additional constraints that were necessary to solve for the field variables that are present in the generalized field equations. These constraint equations are related to response of specific materials. Some thermodynamic inequalities (such as the entropy inequality) were used to find a relationship between the postulated variables that affect the behaviour of the material.

The mechanical behaviour of the material is normally assessed based on mechanical tests such as tension test, shear test, creep test and cyclic fatigue test. It may be noted that a material responds to external fields, in a way that is consistent with its own internal structure. Coupled field responses such as electromechanical response, is also assessed using special tests that are designed to measure specific constants. Details of microstructure of a material are often available only before loading a specimen or after the failure of a specimen. Hence, the idea of linking of the evolution of a microstructure to the continuous deformation of a material, while being a reasonable dream, is not always possible to achieve effectively in actual practice.

Constitutive models are mathematical postulates that relate a set of variables (such as stress) with other set of variables (such as strains) in a material. The validity of these postulates is verifiable only with experimentally observed response of the material. It may be also noted that the material response could be more complicated than the assumptions within which a model is postulated. For example, the stress-strain response of rubber will involve entropic stretching followed by re-crystallization and breaking of bonds. However, the hyperelastic model that attempts to capture the mechanical response, can at best capture the initial stretching but not the hardening behaviour that is due to re-crystallization.

The governing equations such as mass balance, linear momentum balance, angular momentum balance and energy balance are also valid only when there is a continuity of matter within a region. Multiphase systems and discontinuities that are introduced because of disintegration of matter, formation of voids or cracks etc. cannot be analyzed with the balance laws that are developed in Chapter 3. Despite these limitations, continuum mechanics with appropriate constitutive models are used to make a reasonable estimate of the behaviour of such material systems.

The features of the models described in this book and current approaches in developing other models are briefly discussed in Sections 8.2 and 8.3, respectively. A prediction of system behaviour is possible only when the balance laws are solved within the framework of the initial and boundary conditions with which they operate. It is not out of place to take a glimpse of the tools that are used to solve the governing equations in practice, even though a detailed treatment of the same is not part of the scope of this book. This is done in Section 8.4. In Section 8.5, we discuss issues related to understanding the behaviour of overall systems that are designed using engineering materials. Finally, we end with remarks about future challenges.

8.2 FEATURES OF MODELS SUMMARIZED IN THIS BOOK

The book has an introductory chapter (Chapter 1) where the issues associated with modelling are broadly outlined. It was pointed out in that chapter that models can broadly be classified as empirical models, micromechanical models and phenomenological models. It was clarified there that the focus of the current book will be on phenomenological models, which can be used in the broad description of a continuum, within the framework of a subject called *continuum mechanics.* Chapter 2 reviews the basics of tensor algebra, where the basic definitions of tensors, definition of common operators on tensors etc. are recollected. Chapter 3 reviews the basic concepts of continuum mechanics. In this chapter, kinematics of arbitrary motion of particles are developed in detail, the basic laws of motion as applied to a continuum are developed, and constitutive models were introduced as a link between kinetics and kinematics of continuum elements. Some restrictions on the mathematical form of the constitutive equations are also discussed in this context and some simple forms of constitutive equations for thermoelastic materials were developed. Chapter 4 reviews existing linear models that are used by engineers, even without a background of continuum mechanics. Linear elastic, linear viscous and linear viscoelastic models are briefly reviewed in this chapter. Chapter 5 develops the nonlinear fluid models in detail. Non-linear viscous and non-linear viscoelastic models are developed in detail in this chapter and some special features of non-linear response of fluids, such as normal stress differences in the fluids, are introduced and discussed. The chapter ends with a detailed case study of mechanical response of asphalt, where the experimental results on asphalt characterization are presented along some representative models that attempt to capture the experimental results. Chapter 6 reviews the non-linear models in solids. Here, the non-linear elastic response as captured by hyperelastic models, is developed from first principles. Incremental formulations were introduced through hypoelastic models. These incremental formulations were later developed to capture inelastic deformation through the theory of plasticity. The chapter ends with a case study on soil deformation, where the degradation in a clayey soil due to cyclic deformation, was modelled using plasticity theory. Chapter 7 focuses on coupled field models with specific emphasis on shape memory alloys and piezoelectric materials. Modelling of shape memory materials was shown to follow the general principles of plasticity, where the accumulation of plastic strains in the material is monitored using an additional parameter which is sensitive to the evolution of microstructure in the material. Piezoelectric modelling indicated the existence and application of additional balance laws, operating along with the balance laws of mechanics that are already operating on representative element. The presence of additional fields, required the formulation of additional constitutive equations that will have parameters which are indicative of coupling between electrical and mechanical fields that operate on the material. Chapter 8, which is the current chapter, discusses the system response using continuum mechanics and outlines the strengths and limitations of the continuum hypothesis.

8.3 CURRENT APPROACHES FOR CONSTITUTIVE MODELLING

It continues to be a dream of every modeler to make any proposed model physically meaningful. A modeler makes sincere effort to ensure that the proposed mathematical structure, somehow captures the mechanisms that are known to contribute to the overall mechanical behaviour at any given time. Given below are a few attempts that are made to capture these links in different classes of material :

(a) *Mechanisms of ductile deformation in metals*: It is well known that dislocations, their density and their motion is responsible for plastic deformation of metals. Further, it is known that separation of material at grain boundaries in the presence of hard particles, is responsible for fracture in metals. A precise capture of these mechanisms through mathematical equations and incorporation of the same in the overall framework of a constitutive model of a material is a challenge that is the endeavour of researchers working in the area of micromechanics of deformation and fracture of ductile materials. Despite these limitations, attempts to correlate the plastic strain recovery in SMAs, to the relative percentage of solid state fractions (a micromechanical detail) did yield reasonable good results.

(b) *Mechanisms of failure in brittle materials*: Materials like cement, concrete, are a composite consisting of hard particles embedded in brittle matrix. It is known empirically that failure of the material even in compression is associated with formation of micro-cracks and their propagation within the material. Further, it is empirically known that concretes with different compressive strengths, can be made by changing the properties of the matrix, aggregates and their relative percentage within the composite. However, a prediction of the mix behaviour, based on a rudimentary understanding of individual behaviour (during deformation as well as failure), is an issue that is not yet adequately addressed in the research community.

(c) *Mechanisms of deformation and failure of polymers*: A correlation between the elasticity of polymeric networks and their crosslink density, is well known to modelers of polymer/rubber elasticity. Network models of rubber elasticity have captured this correlation in early models for rubber deformation. However, the correlation between known mechanisms of deformation at large strains, such as network breakage, re-crystallization etc. are not yet captured in a mathematical form that can enter into constitutive equations associated with pure mechanical deformation. This is also the reason for the non-availability of a model that can closely link the volume fractions of crystalline and amorphous states of the polymer, to the properties like plastic strain recovery, in shape memory polymers.

(d) *Piezoelectric behaviour of PZT and domain switching*: There are known internal mechanisms such as rotations or switching of crystal lattice that occurs due to the application of an electric field. This domain switching has been identified as the single reason for the piezoelectric effect in these materials. The interaction of grains in which these domains exist, also plays a very important role in the overall piezoelectric properties of these materials. The piezoelectric properties in these materials are typically monitored through curves that relate electric field and polarization (*hysteresis curves*) or through plots that exist between mechanical strains and electric field (*butterfly curves*). Both these curves are highly non-linear curves and hence cannot be directly captured by the simple linear equations that were introduced in Section 6.2. Attempts are made by researchers to capture these curves through a postulated mechanism for domain switching and grain boundary movement at a material level and use these models to predict the response of a structure using finite element analysis.

(e) *Modelling of flowing polymer systems*: During polymer processing operations, molecular orientation and/or crystallization can occur. It is important to analyze these processes, as these affect the response of the polymer. To model these processes, it is important to consider the mechanisms at molecular level and use continuum models to represent these mechanisms. The effect of local stress/strain fields on orientation and crystallization has to be included in the modelling. At the same time, based on the amount and nature of orientation and crystallization, material behaviour is affected. Flow of liquid crystalline polymers also offers similar challenges. Many commercial polymeric systems are blends of two polymers. During polymer processing, phase separation can occur and therefore, the effect of stress/strain fields on phase equilibria has to be modelled. Many modelling efforts postulate an internal variable and describe its evolution along the other balance equations. These internal variables include degree of orientation, crystal volume fraction, phase fraction etc.

(f) *Modelling of colloids, emulsions, granular materials*: Colloidal solutions, emulsions and granular materials are multiphase systems and are used in several engineering applications. It was mentioned in Chapter 1, that balance equations described in this book are applicable within each phase in a multiphase system, and jump conditions have to be formulated for the phase boundaries. However, the preliminary models of mechanical response have been homogenized versions of balance equations described in Chapter 3. The homogenized equations include *effective* material properties that are functions of phase fraction and the resulting equations are similar to single phase equations. For example, based on this type of modelling, we can define an effective viscosity for a colloid or an emulsion.

(g) *Modelling of electro and magneto-rheological fluids*: These are new class of materials that respond to electromagnetic fields. Similar to the discussion in Chapter 7, electro- or magneto-mechanical response of these fluids has to be modelled by considering governing equations of electrodynamics and mechanics. These fluids are usually suspensions of particles in a fluid. When subjected to electrical or magnetic field, particles interact with each other due to their electromagnetic properties. These interactions lead to structural re-arrangement of particles. This structural evolution has to be modelled along with governing equations, if we want to simulate the overall response of systems using these materials.

8.4 NUMERICAL SIMULATION OF SYSTEM RESPONSE USING CONTINUUM MODELS

It may be noted that the available set of equations that need to be solved in continuum mechanics, are a set of first order differential equations as indicated Eqn. (3.3.52) in Chapter 3. It was indicated earlier that additional set of equations that are required to close the set of Eqn. (3.3.52) (in order to make them amenable for solution), are provided by the constitutive relations. It may be noted that governing equations as well as the constitutive equations, in general involve a gradient operation, a divergence operation or a rate operation. While the divergence and gradient operations are operations in space, the rate operation is an operation in time.

If the constitutive equations are explicit, it may be possible to incorporate them into the governing differential equations, so as to arrive at a single equation involving the kinematic variables. In fact, this is normally done for fluid flow using Newtonian fluid or inviscid fluid models. The resulting equations will involve higher order differential terms and analytical solutions may be available for such equations for simple boundary conditions. The Navier Stokes' equation, apart from being a higher order equation, has quite a few non-linear terms, which require special techniques for solution.

It may be noted that as we add more complexity into the constitutive relations, analytical solution of the governing equations will not be possible for all possible boundary conditions. Hence, it will be necessary to utilize numerical approximations to the governing differential equations. Any numerical scheme will focus on converting the given set of differential or integral equations into a set of algebraic equations and solve the algebraic equations for the involved field variables.

The process of converting the differential equations into appropriate algebraic equations is achieved differently for the *temporal variables* (time and frequency) and for spatial variables (x, y and z in rectangular co-ordinates or z, r and θ in cylindrical co-ordinates). Even though the differential equations are likely to have mixed derivatives of temporal and spatial variables, it is customary to treat these two sets of variables as independent variables influencing the field variables. Problems involving temporal variations are classified as *initial value problems* where the initial conditions of the field variables or its time derivative is specified at the start of a deformation process. The solution is marched forward from the solution from one time to another discrete time. The variation of the field variable is normally assumed to be linear between these discrete time steps. The differential equations in time are converted into simple difference equations so that a solution can be obtained at discrete time steps. This scheme of solving for the field variables in time, is called as the *finite difference method* and is widely adopted for solving the equations in fluid mechanics. For some special problems, where the input signal is likely to have signals of wide range of frequencies, it is possible to convert the time domain problems into frequency domain problems by adopting transform techniques like *Laplace transforms* or *Fourier transforms* and solve the problem in the frequency domain. The solutions are later converted back to the time domain by using inverse transforms.

Approximate solutions in spatial domain are obtained by solving the appropriate *boundary value problem* using appropriate numerical scheme. The approximate solutions for the boundary value problems can broadly be stated as methods that try to minimize the error associated with an approximation of the field variable, which is assumed to vary in a particular way within the domain of consideration.The assumed variation, which is normally based on an interpolation between discrete field points, is used to satisfy the given differential equation in space, as well as the boundary conditions. The error associated with the satisfaction of the governing equation in space, is minimized in different ways in different schemes. If the error is reduced to zero only at discrete points in the domain, we have the *collocation procedure* for solving the differential equation. If the error of the governing equation is minimized in an average sense over the domain, using special weighting functions, which are similar to the field interpolating functions, we have the popular version of the *finite element method* to solve the problem. This method is widely used to solve problems in solid mechanics. If the interpolation of the field variable is done locally within a domain and the errors are then minimized, we have the *finite volume method*, which finds an application in many fluid mechanics problems. If the errors within the domain are completely eliminated through the choice of an intelligent interpolating function, and the errors of interpolation are primarily focused on the boundaries, we have the *boundary element method*. This method is popular with potential flow problems in fluid mechanics and elasto-statics problems in solid mechanics.

The variables that are normally solved for in any of the above numerical techniques are either the essential variables or the natural variables, which are defined at discrete points in the domain and at the boundaries. The given differential equations are converted into a system of linear algebraic equations. The coefficients of these equations are known and unknowns that need to be solved for, are either the nodal values of essential variables, natural variables or a combination of both set of variables. The field variables at intermediate points (interior to the nodes), are later evaluated from the nodal point variables, using appropriate interpolation functions.

It is to be noted that many realistic material models that we use in solid mechanics or fluid mechanics, are non-linear in form. In simple terms, this would mean that the relation between the natural variable (like the stress) and the essential variable (like the strain) is not established with a parameter independent, constant coefficient. Hence, the solution of the governing equations that is obtained using these constitutive models, is also not likely to yield a linear variation of the field variable (like the displacement) with the applied loading parameters.

A classical way of obtaining solution for non-linear problems, is to linearize the equations within small intervals and to obtain the solution incrementally for incremental increase in the loading parameters, and to add these incremental solutions to the total solution obtained till the previous time step. It is also customary to use all the information about the problem solution, that is available till the previous time step and to obtain the solution at the current time step. The current solution can further be improved by successive iterations till the improvements obtained reach a saturation limit. One of the popular methods that is used for incremental solution of non-linear equations, is the *Newton-Raphson method*, which works on the principle of using the tangent modulus of the load deflection curve, to obtain the incremental solutions for the governing equations.

8.5 OBSERVATIONS ON SYSTEM RESPONSE

Simulation of a system response based on some understanding of material that constitutes the system, is the aim of any modelling exercise. A quick review of the issues involved in such a modelling exercise, would be appropriate here. These are listed systematically and discussed below:

(*a*) *Observation of material response*: Our understanding of a material behaviour is invariably through mechanical tests on material. Very often, tests are conducted in uniaxial conditions only, while the material models are developed for multiaxial state of stress and calibrated with uniaxial test results. Observations are also primarily made for the essential variables like positions as a function of time and other kinematic variables like displacements, velocities, accelerations, strains are interpolated from these observations. Measurement of kinetic quantities like forces, torques or pressures are often made through transducers, which are once again are primarily sensing displacements and correlating them to forces through an internal calibration.

(*b*) *Modelling of material response*: Many constitutive models, which are developed for materials, satisfy a mathematical structure of symmetry and display their ability to adapt complicated variations, through the use of sufficient number of parameters. Many times, it is not easy to find a physical meaning to the parameters that are suggested by the models or to suggest a test procedure that can estimate a chosen parameter uniquely for a given material. Material models having as many as tens of constants are suggested for the mechanical response of a material and an estimate of these constants uniquely for any new material is not possible..

(c) *Validity of the governing equations*: The equations that we solve for in continuum mechanics, assume a continuity of matter and all the field variables that are associated with the matter. Hence, separation of matter, that is likely to happen locally, as a consequence of the applied loads, cannot be captured by the field equations. However, the consequences of this separation in terms of the overall response of a representative volume element is often captured in the form of equivalent loss of material properties (such as stiffness) and the analysis is still carried out within the continuum hypothesis. Very often, when different sets of equations are combined together to get a single large equation, certain assumptions are made with regard to the nature of kinematic variables or with regard to the nature of constitutive relations. For example, we often find that only the first gradients (strains) are related to stresses in solid mechanics and not their higher powers, for most solids. Further, this relationship is often assumed to be in terms of a single, parameter independent constant in most theories of linearized elasticity. It is to be noted that this assumption will not hold true for all materials under all loading situations. Hence, the solutions that are obtained using such approaches need always to be interpreted within the constraints of these assumptions.

(d) *Validity of observations of system response*: Material models are often developed as inputs into a system of equations that govern the behaviour of a system. The validity of the material model and its integrity with the governing equations can finally be verified only when we make observations on a system. Hence, there is a necessity to validate the modelling philosophy with some observations at a system level. Observations that can be made at a system level are often more gross compared to the observations that can be made for material characterization. For example, observations that can be made in the mechanical performance of a rubber tyre are different from the observations that can made to characterize the material that constitute a tyre structure. It may be possible to measure the radial displacement or hoop expansion of a tyre. If our material model is correct, our governing equations are valid and if the geometry of the tyre is accurately modelled, then the observed system response must be predicted fairly accurately by our system model. It may be noted that a system response in one setting may also be predicted by an alternate material model (often simpler) that is assumed to operate on a simplified idealization of the system. In this example, a tyre can be modelled as a linear spring having an equivalent stiffness, even without worrying too much about the non-linear constitutive relation of rubber. However, the predictions that can be made using such crude models (both of the system as well as the material) will always be limited.

(e) *Predictions based on the model developed*: Material models and the predictions and interpretations, that are made using material models are relevant for predictions of field situations where one has no physical access to make physical measurements. The response of a new multi-storey building, the mechanical response of a nuclear vessel when subjected to thermal cycles, the response of clayey soils when they are subjected to earthquake loading, vibrations of an aircraft when it is going through turbulent winds etc., are difficult to measure when the system is in operation. The responses of such system are often predicted using system models. In all these cases, the confidence of our predictive capability will depend on the choice of the appropriate material model, the use of the right governing equations, the accuracy of the geometric models to capture the overall geometry and the boundary conditions, the accuracy of the numerical scheme that is used to solve the algebraic equations associated with the system response etc. Hence, the use of correct material model is an important ingredient in predicting system response in all such complicated systems.

8.6 CHALLENGES FOR THE FUTURE

Great advances are taking place in understanding of material structure at microscopic, molecular and atomic level through experimental techniques of microscopy. Similarly, microscopic, molecular and atomistic modelling has also led to insights into material response at those levels. The greatest challenge lies in using information from these understandings/insights into models which can be utilized in analyzing system response in an engineering application. Therefore, continuum models are being refined by incorporating many features representing microscopic/molecular/atomistic mechanisms.This exchange of ideas and information across levels of materials, is being attempted in *multiscale modelling* or *micro-macro approach*.

Engineers will continue to use materials or operate with materials in challenging environments. The scale of operations of these materials range from massive structures operating in difficult environment (such as nuclear structures operating at high temperatures), to the small scale operations that are even difficult to observe and monitor (such as the flow of complex fluids like blood flowing through minute arteries). In all these cases, engineers continue to design and build their systems based on a rudimentary behaviour of the system behaviour. The lack of complete understanding of any system has never been an impediment for the design and execution of a system. At any stage, some rudimentary models of the system behaviour are used to make an intuitive guess of system behaviour and a risk is taken to deploy the system and to monitor its behaviour.

Attempts to capture system behaviour within the framework of known laws of physics and mathematics are always done later, with a hope that the known laws of physics and mathematics can reasonably predict all the phenomena that take place in the world. Material modelling is a component of the system modelling, where it is hoped that all observed complications in a material, in association with its known causes, can be captured in a reasonable way within a mathematical framework, so as to be useful in the prediction of system behaviour somewhere. Capturing a known experimental observation within the framework of a mathematical structure, by itself seems to be a formidable task that occupies the attention of many researchers working in continuum mechanics. These researchers demonstrate the working of their sophisticated model in simple loading conditions. Researchers working with the system simulation find that the system integration, involving a complex interplay between the various components is so complicated that they really do not mind some inaccuracies in the mathematical structure of material model associated with individual components. Hence, the development of material models that are not only realistic in their description, but also will be simple enough to use in system prediction, will continue to be a challenge for researchers in future.

■■■

Appendix
General and Convected
Coordinates

Problem formulation in rectangular, spherical or cylindrical coordinates is adequate for most of the engineering problems. We approximate real geometries to be idealizations represented by one of the above three coordinate systems. These coordinates are also used so often because we can visualize the problems with relative ease. Of these, cylindrical and spherical are curvilinear, though all of them are orthogonal coordinates.

The system model and the material model should be valid for all possible coordinate systems. Therefore, generalized coordinates are used to develop the governing equations and the constitutive models. The generalized coordinate system implies that it may be neither orthogonal nor rectilinear. In the context of mechanics of materials, we can think of coordinates which are not fixed but moving, deforming or rotating with the material itself. Since we require the material behaviour to conform to general physical notions such as frame indifference, we will see that governing descriptions should be related to kinematics of material. Since the material may be subjected to arbitrary translation, deformation and rotation, we cannot assume orthogonal or rectilinear coordinate system.

As stated earlier, for applications, we may not use the non-orthogonal coordinates. So, governing equations developed by using generalized coordinate systems are transformed to rectangular, cylindrical or spherical coordinates as the need may be.

Let us look closely at the cylindrical curvilinear coordinate system. To consider the three dimensional space, we require r, q and z coordinates. The three base vectors required to describe the space can be thought to arise out of two considerations. Let us consider two neighbouring points at **r** and **r** + d**r**, as shown in Fig. A1.

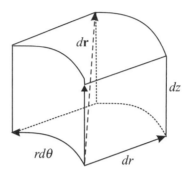

Fig. A1 Elemental volume in cylindrical coordinates

Vector joining these two points is d**r**. The line elements in each of the three directions (dr, $rd\theta$ and dz) are related to d**r** as

$$d\mathbf{r} = dr\mathbf{e}_r + rd\theta\mathbf{e}_\theta + dz\mathbf{e}_z \qquad (A.1)$$

We can define the direction in which θ and z are constant as the direction of one of the base vectors as shown in Fig. A2. Similarly, two other base vectors can be defined to be in directions along constant r & θ and constant r & z, respectively.

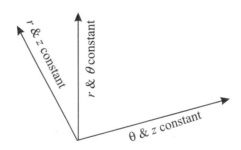

Fig. A2 Directions showing constant coordinates

Alternately, we can define base vectors such that they are perpendicular to constant parameter surface. For example, we can define a base vector as normal to the surface on which r is constant. Similarly, two other base vectors can defined as normal vectors to θ-constant and z-constant surfaces, respectively. These base vectors are shown in Fig. A3.

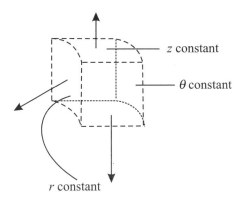

Fig. A3 Directions perpendicular to constant coordinate surfaces

We can understand the above two descriptions through following two simplifying statements:

(*i*) θ and z constant → what is the change with respect to r?

(*ii*) perpendicular to r-constant surface → what is the gradient of r?

In both the above cases, we get a base vector related to r coordinate. Let us see how we can define general coordinates based on these two formulations. We will define the base vectors in terms of the known cylindrical coordinate system ($\mathbf{e}_r$, $\mathbf{e}_\theta$, $\mathbf{e}_z$). We will denote the generalized base vectors as ($\mathbf{g}_r$, $\mathbf{g}_\theta$, $\mathbf{g}_z$) and ($\mathbf{g}^r$, $\mathbf{g}^\theta$, $\mathbf{g}^z$).

Example A1 : Write the cylindrical coordinates in terms of generalized coordinates.

Solution :

As shown in Fig. AEI, $\mathbf{r}_p$ is the position vector, denoted by $r\mathbf{e}_r + z\mathbf{e}_z$ in cylindrical coordinates.

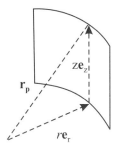

Fig. AEI Position vector as addition of two vectors along radial and axial directions, respectively

General coordinates based on measures of r, θ and z :

1. What is the change in $\mathbf{r}_p$ with r, θ, and z (or (θ, z) constant, (z, r) constant, (r, θ) constant)? Based on this question, we can write

$$\mathbf{g}_r = \frac{\partial \mathbf{r}_p}{\partial r} = \mathbf{e}_r$$

$$\mathbf{g}_\theta = \frac{\partial \mathbf{r}_p}{\partial \theta} = r\mathbf{e}_\theta \qquad\qquad (A.E.1)$$

$$\mathbf{g}_z = \frac{\partial \mathbf{r}_p}{\partial z} = \mathbf{e}_z$$

2. What are the gradients of r, θ, and z? Based on this question, we can write

$$\mathbf{g}^r = \operatorname{grad} r = \mathbf{e}_r$$

$$\mathbf{g}^\theta = \operatorname{grad} \theta = \frac{1}{r}\mathbf{e}_\theta \qquad\qquad (A.E.2)$$

$$\mathbf{g}^z = \operatorname{grad} z = \mathbf{e}_z$$

$\mathbf{g}^r$, $\mathbf{g}^\theta$ and $\mathbf{g}^z$ are called the contravariant base vectors and $\mathbf{g}_r$, $\mathbf{g}_\theta$ and $\mathbf{g}_z$ are called the covariant base vectors.

Notice that the units of the generalized base vectors in Example A1 are not all the same. They are also not unit vectors. Any arbitrary vector can be expressed in terms of them as

$$\mathbf{v} = v_r\mathbf{g}^r + v_q\mathbf{g}^\theta + v_z\mathbf{g}^z \qquad\qquad (A.2)$$

$$\mathbf{v} = v^r\mathbf{g}_r + v^\theta\mathbf{g}_\theta + v^z\mathbf{g}_z \qquad\qquad (A.3)$$

where v_r, v_θ and v_z are called the contravariant components and v^r, v^θ and v^z are called the covariant components. Again, note that the units of v_i and v^i may not be the same. We can also define operators based on these base vectors, such as the gradient operator

$$\operatorname{grad} = \mathbf{g}^r\frac{\partial}{\partial r} + \mathbf{g}^\theta\frac{\partial}{\partial \theta} + \mathbf{g}^z\frac{\partial}{\partial z} \qquad\qquad (A.4)$$

In the case of generalized coordinates based on the cylindrical geometry, the base vectors remain in the same direction as the familiar unit vectors $(\mathbf{e}_r, \mathbf{e}_\theta, \mathbf{e}_z)$. In fact, you can show that for a rectangular coordinate system, the two sets of base vectors are identical in direction as well as magnitude. Therefore, there is no specific advantage in using either the contravariant or the covariant bases. Instead, we define a fixed cylindrical coordinate system with unit vectors as the base vectors. In general, however, we have two choices in terms of description, covariant description (base vectors) or contravariant description.

Let us define the general curvilinear coordinate and the base vectors associated with them. Unlike in the above example, these are not dependent on any specific geometry or material deformation. After introducing these general coordinates, we will introduce the convected coordinates, which are associated with deformation in the material. We will denote the general coordinates as q^1, q^2 and q^3. As shown in Fig. A4, we can define the two sets of base vectors.

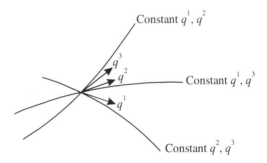

Fig. A4 Definition of generalized coordinates

Similar to the Example A1, we can write the base vectors and a line element as,

$$\mathbf{g}_i = \frac{\partial \mathbf{r}}{\partial q^i}$$
$$d\mathbf{r} = dq^i \mathbf{g}_i \tag{A.5}$$

and

$$\mathbf{g}^i = \text{grad } q^i$$
$$d\mathbf{r} = dq^i \mathbf{g}^i \tag{A.6}$$

For simplistic problems, we use the governing equations in orthogonal fixed coordinate systems, namely rectangular, cylindrical and spherical. However, in mechanics of materials, it is very useful to also consider material behaviour in coordinate systems which are moving along with the material. We keep track of kinematics in order to understand material motion. If the coordinate system is inherently related to material motion, then kinematic quantities will be related to coordinate measures.

The advantage of frame and coordinates embedded in the material is that they are locally defined and the relationships proposed thus have shown to be frame indifferent and not dependent on the coordinate systems. Few of the popular coordinates used in engineering applications are corotating, codeformational or convected coordinates.

Let us consider convected coordinates, which are embedded in the material and deform along with the material. Let us denote the convected coordinates by y_i and the base vectors by $\mathbf{g}_i$ and $\mathbf{g}^i$. We choose a fixed frame with a coordinate system such that at present time it coincides with the convected coordinates. So that, at present time:

$$\mathbf{g}_i = \mathbf{g}^i = \mathbf{e}_i \text{ and } q^i = x_i \quad \text{at time } t \qquad (A.7)$$

However, at any arbitrary time τ, a material point has configuration denoted by x_i^τ. However, a material point will continue to have the same q_i that it has at present time, since the q_i are convected coordinates and they are moving along with the material. In other words, q_i are the labels of a material particle. We know that based on kinematical quantities $\mathbf{x}^\tau$ is related to $\mathbf{x}$. Configuration of a particle at time t can be either expressed in terms of convected coordinate (or its label) or in terms of the present configuration

$$x_i^\tau = x_i^\tau\left(q^j, \tau\right)$$

$$= x_i^\tau\left(x_j, t, \tau\right) \qquad \text{... (A.8)}$$

A differential line element and its length in the convected coordinate is given by :

$$d\mathbf{r} = \mathbf{g}_i dq^i$$

$$d\mathbf{r} . d\mathbf{r} = \mathbf{g}_i \cdot \mathbf{g}_j \, dq^i dq^j \qquad \text{... (A.9)}$$

where $\mathbf{g}_i \cdot \mathbf{g}_j = g_{ij}$ are the metric coefficients. The metric coefficients contain all the information about the deformation in the material. The convected coordinates are embedded in the material and change with reference to a fixed reference frame. Therefore, the difference in metric coefficients at two different instants of time gives an idea about the deformation experienced by a differential line element. In other words, the deformation measures can be defined on the basis of the base vectors and the metric coefficients.

In mechanics, we are interested in deformation, and hence we are interested in keeping track of relative distance between two material points. Measures of distance would form a natural choice to be involved in strain measures.

Example A2: Express metric coefficients in rectangular and cylindrical coordinates.

Solution:

Let us look at the distance measures between two neighbourhood points. If we choose cartesian coordinate system, we can write

$$ds^2 = d\mathbf{r} . d\mathbf{r} = dx^2 + dy^2 + dz^2$$

or

$$ds^2 = dx_i . dx_i = \delta_{ij} dx_i dx_j \qquad (A.E.3)$$

If we choose another coordinate system to describe the two points, we can write

$$x_i = x_i(y_i) \tag{A.E.4}$$

where $dx_p = \dfrac{dx_p}{dy_i} dy_i$ and we can write the distance measure as

$$ds^2 = \frac{\partial x_m}{\partial y_i} \frac{\partial x_m}{\partial y_j} \partial y_i \partial y_j \tag{A.E.5}$$

$\dfrac{\partial x_m}{\partial y_i} \dfrac{\partial x_m}{\partial y_j}$ are called the metric coefficients. If y_i are cylindrical coordinates, we have

$$y_1, y_2, y_3 \rightarrow r, \theta, z$$
$$x_1, x_2, x_3 \rightarrow x, y, z \tag{A.E.6}$$

Therefore, we can estimate the metric coefficients to be

$$\frac{\partial x_m}{\partial y_i} \frac{\partial x_m}{\partial y_j} = \begin{bmatrix} 1 & 0 & 0 \\ 0 & r^2 & 0 \\ 0 & 0 & 1 \end{bmatrix}$$

This is due to,

$$x_i = x_i(y_i) \Rightarrow x = r\cos\theta, \ y = r\sin\theta, \ z = z$$

$$i = 1, j = 1 \Rightarrow \frac{\partial x_m}{\partial r} \frac{\partial x_m}{\partial r} = \left(\frac{\partial x}{\partial r}\right)^2 + \left(\frac{\partial y}{\partial r}\right)^2 + \left(\frac{\partial z}{\partial r}\right)^2$$

$$= \cos^2\theta + \sin^2\theta = 1 \tag{A.E.8}$$

$$i = 1, j = 2 \Rightarrow \frac{\partial x_m}{\partial r} \frac{\partial x_m}{\partial \theta}$$

$$= \cos\theta(-r\sin\theta) + \sin\theta\, r\sin\theta$$

So that,

$$ds^2 = dr^2 + r^2 (d\theta)^2 + dz^2 \tag{A.E.9}$$

As explained earlier, a vector or a tensor can be represented in the convected coordinates. For example, traction on surfaces characterized by base vectors will be as follows

$$\mathbf{t}_i = \frac{\mathbf{g}_i}{|\mathbf{g}_i|} \cdot \boldsymbol{\sigma} \tag{A.E.10}$$

The component of the traction along a base vector will be given by,

$$\mathbf{t}_i \cdot \mathbf{g}_j = \left[\frac{\mathbf{g}_i}{|\mathbf{g}_i|} \cdot \boldsymbol{\sigma}\right] \cdot \mathbf{g}_j = \frac{1}{|\mathbf{g}_i|} \boldsymbol{\sigma} : \mathbf{g}_j \mathbf{g}_i \qquad \dots \text{(A.E.11)}$$

The quantity $\boldsymbol{\sigma} : \mathbf{g}_j \mathbf{g}_i$ is nothing but the component of stress tensor in convected coordinates. (recall how one can find stress component in rectangular coordinate system). Therefore, one can write relations between these stress components and deformation measures, which are also expressed in terms of convected coordinates. Moreover, if these relations involve rate of stress and deformation, we can include those in convected coordinates as well. Relations written in this form will be frame indifferent.

EXERCISE

A.1 Consider one dimensional simple shear flow between two parallel plates. The flow can be described using a rectangular coordinate system (x_1, x_2), which is fixed. The velocity profile in such a coordinate system is given by $v_1 = \dot{\gamma} x_2$ or $\mathbf{v} = \dot{\gamma} x_2 \mathbf{e}_1$

(a) Show that the material particle which is at (x_1, x_2) at time t will be at $[x^\tau_1 = x_1 - \dot{\gamma}(t - \tau)x_2, \; x^\tau_2 = x_2]$ at time t.

(b) Assume that the convected coordinates coincide with the fixed rectangular coordinate system at time t. Therefore, $q_1 = x_1$ and $q_2 = x_2$, at time t. Using definitions of the convected base vectors in Eqns. (A.5) and (A.6), show that

$$\mathbf{g}_1 = \mathbf{e}_1$$

$$\mathbf{g}_2 = -\dot{\gamma}(t - \tau)\mathbf{e}_1 + \mathbf{e}_2$$

$$\mathbf{g}^1 = \mathbf{e}_1 + \dot{\gamma}(t - \tau)\mathbf{e}_2$$

$$\mathbf{g}^2 = \mathbf{e}_2$$

■■■

Bibliography

Bird, R. B. and Hassager, H., *Dynamics of Polymeric Liquids*, Vol. 1, Fluid Mechanics, Wiley-Interscience; 2nd edition, 1987.

Chadwick P., *Continuum Mechanics: Concise Theory and Problems*, Dover, 2nd edition, 1999.

Desai C.S. and Siriwardane H.J., *Constitutive Laws for Engineering Materials – With Emphasis on Geologic Materials*, Prentice-Hall Inc., Englewood Cliffs, New Jersey, 1984.

Haupt, P., *Continuum Mechanics and Theory of Materials*, Springer Verlag, 2nd edition, 2002.

Lagoudas, D.C., *Shape Memory Alloys: Modelling and Engineering Applications*, Springer, 2008.

Larson, R.G., *The Structure and Rheology Complex Fluids*, Oxford University Press, 1999.

Maugin, G.A., *Nonlinear Electromechanical Effects and Applications*, World Scientific Publishing Co Pvt. Ltd., 1985.

Otsuka K. and Wayman, C. M., Editors, *Shape Memory Materials*, Cambridge University Press, 2008.

Pandit, D., *Modelling of Shape Memory Polymers behaviour*, Thesis submitted for the award of Master of Technology, Department of Applied Mechanics, IIT Madras, 2008.

Sathiyanarayanan, S., Electro-mechanical Characterization of PVDF, Thesis submitted for the award of Doctor of Philosophy, Department of Applied Mechanics, IIT Madras, 2005.

Reddy, J.N., *An Introduction to Continuum Mechanics*, Cambridge University Press, 2007.

Sokolnikoff, I.S., *Tensor Analysis: Theory and Applications to Geometry and Mechanics of Continua*, John Wiley; 2nd edition 1964.

Sridhanya, K.V., Investigations on the Cyclic Behavior of Soft Clays, Thesis submitted for the award of Master of Science by Research, Department of Civil Engineering, IIT Madras, 2007.

Treloar, L.R.G., *The Physics of Rubber Elasticity*, Clarendon Press, Oxford, 1975.

Truesdell, C., *The Non-Linear Field Theories of Mechanics*, Springer Verlag, 1965.

Truesdell, C. and Rajagopal, K. R., *An Introduction to the Mechanics of Fluids*, Birkhäuser, 2000.

Wilmński, K., *Thermomechanics of Continua*, Springer Verlag, 1998.

Wineman A.S., Rajagopal, K.R., *Mechanical Response of Polymers: An Introduction*, Cambridge University Press, 2000.

■■■

Index

■■■

विद्या ददाति विनयं विनयात् याति पात्रताम् ।
पात्रत्वात् धानमाप्रोति धानात् धार्मं ततः सुखम ॥
(सुभाषितम्)

vidyā   dadāti vinayam vinayāt yāti pātratām.
pātratvāt dhanamāpnoti dhanāt dharmaṃ
tataḥ sukham. (Subhāṣhitam)

Knowledge gives humility, humility begets
maturity. Maturity begets wealth. Wealth
(earned in this way) establishes order and
yields happiness.